INTERNATIONAL CENTRE FOR MECHANICAL SCIENCES

COURSES AND LECTURES No. 151

MICROPOLAR ELASTICITY

EDITED BY W. NOWACKI AND W. OLSZAK

SYMPOSIUM
ORGANIZED BY THE DEPARTMENT
OF MECHANICS OF SOLIDS
JUNE 1972

UDINE 1974

SPRINGER-VERLAG WIEN GMBH

Originally published by Springer-Verlag Wien-New York in 1972

ISBN 978-3-211-81262-4 ISBN 978-3-7091-2920-3 (eBook)

DOI 10.1007/978-3-7091-2920-3

PREFACE

The theory of asymmetric elasticity has been considerably developed in the last years. This concerns the general theory of Cosserat media with six degrees of freedom, the theory called that of couple stresses with constrained rotations as well as the theory based on the notion of directors in relation to more general polar continua. Analogous developments are being observed in the domain of generalized thermal and distortion (dislocations and disclination) problems when passing from the classical to Cosserat media.

In June-July 1970, following a proposal by A.C. Eringen, W. Nowacki, W. Olszak, and R. Stojanović, a series of Summer Courses was organized at the seat of CISM in Udine; this was devoted to Cosserat problems and their generalizations. In several monographic courses, the participants were presented the linear Cosserat theory; lectures, as well as seminars and discussion were held. These courses, which were attended by young research workers from various European universities, were very successful. They awoke a real interest, and stimulated creative research activities in the domain of polar media. The participants suggested to meet again at a Symposium to be held in two years in order to discuss new trends of development in the field of asymmetric elasticity.

As a result, the above mentioned group of scientists decided to organize, in Udine, in the period June 19th to 23rd, 1972, a Symposium on "Micropolar Elasticity"; specialists in this domain as well as the participants of the 1970 course were invited to take an active part in this meeting. The main idea was to present in a few general reports the present state of the art in the field as well as a critical review of the various directions in which the theory of micropolar media has recently been developing together with the prevailing trends of its progress. Five such general reports have been invited. They were focused on the following subjects: "Linear theory of micropolar elasticity" (W. Nowacki); "Linear micropolar media with constrained rotations" (G. Grioli); "Nonlinear micropolar elasticity" (R. Stojanović); "Continuum and lattice theories of elastic and dielectric solids" (R.D. Mindlin); and "Discrete micropolar media" (C. Woźniak). Unfortunately, R.D. Mindlin prevented by illness, was not able to present his report; this has been replaced by another general report on the "Micropolar thermoelasticity" (W. Nowacki).

Thus, most of the current research field could be covered by the general reports presented. The report on "Nonlinear micropolar elasticity" prepared by R. Stojanović just before his tragic death has been read by his co-workers R. Jarić and D. Ružić.

The programme of the Symposium was scheduled in such a manner that every day one general report was presented. Afterwards original contributions in the corresponding special fields followed, delivered by some of the participants and advancing many original ideas and new results. Finally plenty of time was reserved for discussions. These were very vivid and stimulating, and referred to the general reports as well as to the individual contributions.

The last afternoon was devoted to a round table discussion in which mainly the general trends of development of the micropolar approach were criticially reviewed and treated.

The participants stressed the great efforts being made in the domain of developing the foundations of asymmetric elasticity and the treatment of ensuing theoretical problems; they also stressed the necessity of conceiving and of conducting experimental research programmes, especially in view of determining the relevant material characteristics and constants.

The present volume contains the text of the five general reports presented at the Symposium. We hope it will constitute a useful source of information on the problems discussed during the Symposium.

W. Nowacki *W. Olszak*

Udine, December 1973

THE SYMPOSIUM PROGRAMME

Monday, June 19th, 1972

W. Nowacki: "The Linear Theory of Micropolar Elasticity" (general report).
Individual contributions; discussion.

Tuesday, June 20th, 1972

G. Grioli: "Nonlinear Micropolar Media with Constraint Rotations" (general report).
Individual contributions; discussion.

Wednesday, June 21st, 1972

R. Stojanović: "Nonlinear Micropolar Elasticity" (general report, presented by J. Jarić and D. Ruzić).
Individual contributions; discussion.

Thursday, June 22nd, 1972

W. Nowacki: "Micropolar Thermoelasticity" (general report).
Individual contributions; discussion.

Friday, June 23rd, 1972

C. Woźniak: "Discrete Micropolar Media" (general report).
Individual contributions; discussion.
General round table discussion.

INDIVIDUAL CONTRIBUTIONS

J.B. Alblas (Eindhoven), G. Anderson (Troy, New York), M. Aron (Iași), W. Bürger (Darmstadt), G. Capriz (Pisa), M. Kleiber (Warsaw), Z. Olesiak (Warsaw), A. Paglietti (Cagliari), N. Petrov (Sofia), C. Rymarz (Warsaw).

LIST OF PARTICIPANTS

J.B. Alblas (Eindhoven), G. Anderson (Troy, New York), M. Aron (Iași), J. Bejda (Warsaw), W. Bürger (Darmstadt), G. Capriz (Pisa), A. Cardon (Brussels), I. Deutsch (Brașov), U. Gamer (Vienna), J. Gioanni (Marseille), G. Grioli (Padua), F. Jaburek (Leoben), J. Jarić (Belgrade), C.M. Jeșan (Iasi), J. Kasperkiewicz (Warsaw), G. Kiesselbach (Vienna), M. Kleiber (Warsaw), J. König (Warsaw), N. Konstantinov (Sofia), W. Krzyś (Cracow), G. Lebon (Liège), M. Maniacco (Trieste), M. Mele (Trieste), S. Milanović (Belgrade), H. Nguyen (Liège), J.P. Nowacki (Warsaw), W. Nowacki (Warsaw), Z. Olesiak (Warsaw), J. Orkisz (Cracow), A. Paglietti (Cagliari), N. Petrov (Sofia), M. Plavšić (Belgrade), P. Podio Guidugli (Pisa), T. Ruggeri (Messina), D. Ružić (Belgrade), C. Rymarz (Warsaw), C. Silli (Pisa), S. Slavtchev (Sofia), W. Weigert (Vienna).

CONTENTS

W. NOWACKI

THE LINEAR THEORY OF MICROPOLAR ELASTICITY

1. Introduction

The classical theory of elasticity describes well the behaviour of construction materials (various sorts of steel, aluminium, concrete) provided the stresses do not exceed the elastic limit and no stress concentration occurs.

The discrepancy between the results of the classical theory of elasticity and the experiments appears in all the cases when the microstructure of the body is significant, i. e. in the neighbourhood of the cracks and notches where the stress gradients are considerable. The discrepancies also appear in granular media and multimolecular bodies such as polimers.

The influence of the microstructure is particularly evident in the case of elastic vibrations of high frequency and small wave length.

W. Voigt tried to remove the shortcomings of the classical theory of elasticity [1] by the assumption that the interaction of two parts of the body is transmitted through an area element dA by means not only of the force vector $\underline{p}dA$ but also by the moment vector $\underline{m}dA$. Thus, besides the force stresses σ_{ji} also the moment stresses have been defined.

However, the complete theory of asymmetric elasticity was developed by the brothers François and Eugène Cosserat [2] who published it in 1909 in the work "Théorie des corps deformables".

They assumed that the body consists of interconnected particles in the form of small rigid bodies. During the deformation each particle is displaced by $\underline{u}(\underline{x},t)$ and rotated by $\underline{\varphi}(\underline{x},t)$, the functions of the position $\underline{x}$ and time t.

Thus an elastic continuum has been described such that its points possess the orientation (polar media) and for which we can speak of the rotation of a point. The vectors $\underline{u}$ and $\underline{\varphi}$ are mutually independent and determine the deformation of the body. The introduction of the vectors $\underline{u}$ and $\underline{\varphi}$ and the assumption that the transmission of forces through an area element dA is carried out by means of the force vector $\underline{p}$ and the moment vector $\underline{m}$ leads in the consequence to asymmetric stress tensors σ_{ji} and μ_{ji}.

The theory of the brothers E. and F. Cosserat remained unnoticed and was not duly appreciated during their lifetime. This was so because the presentation was very general (the theory was non-linear, including large deformations) and because its frames exceeded the frames of the theory of elasticity. They attempted to construct the unified field theory, containing mechanics, optics and electrodynamics and combined by a general principle of the least action.

The research in the field of the general theories of continuous media conducted in the last fifteen years, drew the attention of the scientists to Cosserats' work. Looking for the new models, describing more precisely the behaviour of the real elastic media, the models similar to, or identical with that of Cosserats' have been encountered. Here, we mention, first of all, the papers by C. Truesdell and R. A. Toupin [3], G. Grioli [4], R. D. Mindlin and H. F. Tiersten [5]. At the beginning the author's attention was concentrated on the simplified theory of elasticity, so called the Cosserat pseudo-continuum. By this name we understand a continuum for which the asymmetric force stresses and moment stresses occur, however, the deformation is determined by the displacement vector $\underline{u}$ only. Here we assume, as in the classical theory of elasticity, that $\underset{\sim}{\varphi} = \frac{1}{2} \operatorname{curl} \underline{u}$. It is interesting to notice that this model was also considered by the Cosserats who called it the model with the latent trihedron.

A number of German authors, W. Günther, H. Schäfer [7], H. Neuber [8] referred directly to the general theory of Cosserats supplementing it with constitutive equations. The general relations and equations of the Cosserats' theory have also been derived by E. V. Kuvshinskii and A. L. Aero [9] and N. A. Palmov [10]. Here one should also mention the generalizing work by A. C. Eringen and E. S. Suhubi [11].

At the present moment the theory of Cosserats is in the full development. The literature on the subject increases, and the problems of the asymmetric theory of elasticity were discussed in two symposia, namely IUTAM Symposium in Freudenstadt in 1968 and in this Symposium organized by CISM. Likewise the first monographs devoted to the micropolar elasticity, by R. Stojanovic [12] and W. Nowacki [13] appeared, both were published in 1970.

The discussion in the present work is confined to the linear theory of the micropolar elasticity. We begin with the dynamic problems, then we consider the statical ones.

2. The Dynamical Problems of the Micropolar Elasticity

Let us consider a regular region $V+A$ bounded by a smooth surface A, containing a homogeneous, isotropic, centrosymmetric and micropolar continuum of the density ϱ and the rotational inertia $\mathfrak{J}$.

The body is deformed by the external loading. Let on part A_σ of the bounding surface of the body the forces $\underline{p}$ and the moments $\underline{m}$ act, while on part A_u the rotations $\underline{\varphi}$

and displacements $\underline{u}$ be prescribed. The body forces $\underline{X}$ and the body moments $\underline{Y}$ act inside the body. The loadings generate the deformation of the body described by the displacement vector $\underline{u}(\underline{x},t)$ and the rotation vector $\varphi(\underline{x},t)$. Consequently, in the body there develop the force stresses $\sigma_{ji}(\underline{x},t)$ and the moment stresses $\mu_{ji}(\underline{x},t)$. The components σ_{1i}, μ_{1i} of these stresses are presented in Fig. 1. The stresses σ_{ji}, μ_{ji} are connected with the asymmetric tensor of deformation γ_{ji} and the torsion flexure tensor $\varkappa_{ji}$. The dynamic problem of the micropolar theory of elasticity consists in determining the stresses σ_{ji}, μ_{ji} the deformations γ_{ji}, $\varkappa_{ji}$ the displacement $\underline{u}$, and the rotation $\underline{\varphi}$. These functions should satisfy the equations of motion, the constitutive equations, the boundary conditions, and the initial conditions.

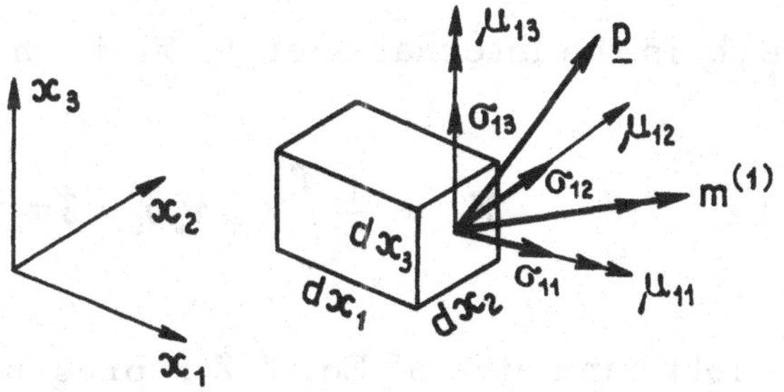

$$\underline{p}^{(1)} = (\sigma_{11}, \sigma_{12}, \sigma_{13})$$

$$m^{(1)} = (\mu_{11}, \mu_{12}, \mu_{13})$$

Fig. 1

The equations of motion take the form

$$\sigma_{ji,j} + X_i = \rho \ddot{u}_i \quad ,$$
$$\epsilon_{ijk}\sigma_{jk} + \mu_{ji,j} + Y_i = J\ddot{\varphi}_i \quad . \qquad (2.1)$$

In these equations, written in the cartesian orthogonal coordinate system, ϵ_{ijk} is Ricci's alternator, ρ denotes the density, while $\mathcal{J}$ is the rotational inertia.

The constitutive equations can be obtained from the following discussion. We have, from the principle of the energy conservation, under the assumption of an adiabatic process that

$$(2.2)\qquad \frac{d}{dt}(\mathcal{U}+\mathcal{K}) = \int_V (X_i v_i + Y_i w_i)\,dV + \int_A (p_i v_i + m_i v_i)\,dA\,, \quad v_i = \dot{u}_i\,, \; w_i = \dot{\varphi}_i\,.$$

Here $\mathcal{U}$ is the internal energy, $\mathcal{K}$ is the kinetic energy where

$$(2.3)\qquad \mathcal{K} = \frac{1}{2}\int_V (\rho\, v_i v_i + \mathcal{J} w_i w_i)\,dV\,.$$

The right hand side of Eq. 2.2 represents the power of the external forces. Taking into account the equations of motion (1.1) we obtain

$$(2.4)\qquad \dot{U} = \sigma_{ji}\dot{\gamma}_{ji} + \mu_{ji}\dot{\varkappa}_{ji}\,, \qquad \mathcal{U} = \int_V U\,dV\,, \qquad U = U(\gamma_{ji}, \varkappa_{ji})\,.$$

Hence we obtain the definition of the deformation tensors

$$(2.5)\qquad \gamma_{ji} = u_{i,j} - \epsilon_{kji}\varphi_k\,, \qquad \varkappa_{ji} = \varphi_{i,j}\,.$$

The internal energy U is the function of independent variables γ_{ji}, $\varkappa_{ji}$ and is the function of state. Thus we have

$$(2.6)\qquad \dot{U} = \frac{\partial U}{\partial \gamma_{ji}}\dot{\gamma}_{ji} + \frac{\partial U}{\partial \varkappa_{ji}}\dot{\varkappa}_{ji}\,.$$

We assume that the functions σ_{ji}, μ_{ji} do not depend explicitly on the time derivatives of the functions γ_{ji}, $\varkappa_{ji}$. We have

$$\sigma_{ji} = \frac{\partial U}{\partial \gamma_{ji}} , \quad \mu_{ji} = \frac{\partial U}{\partial \varkappa_{ji}} . \tag{2.7}$$

The internal energy can be represented in the following form

$$U = \frac{\mu+\alpha}{2}\gamma_{ji}\gamma_{ji} + \frac{\mu-\alpha}{2}\gamma_{ji}\gamma_{ij} + \frac{1}{2}\lambda\gamma_{kk}\gamma_{nn} + \frac{\gamma+\varepsilon}{2}\varkappa_{ji}\varkappa_{ji} + \frac{\gamma-\varepsilon}{2}\varkappa_{ji}\varkappa_{ij} + \frac{\beta}{2}\varkappa_{kk}\varkappa_{nn} . \tag{2.8}$$

The form of the energy, presented here, can be justified in the following way. Since the internal energy is scalar, then each term on the right hand side of the equation must also be a scalar. By means of the components of the tensor γ_{ji} one can construct three independent square invariants, namely $\gamma_{ji}\gamma_{ji}$, $\gamma_{ji}\gamma_{ij}$ and $\gamma_{kk}\gamma_{nn}$. The same thing refers to the tensor $\varkappa_{ji}$. The terms $\gamma_{ji}\varkappa_{ji}$, $\gamma_{ji}\varkappa_{ij}$ and $\gamma_{kk}\varkappa_{nn}$ do not enter the expression (2. 8) since this would contradict the postulate of the centrosymmetry. Thus, we have six material constants $\mu, \lambda, \alpha, \beta$, measured in the adiabatic conditions. These constants should satisfy the following inequalities

$$3\lambda + 2\mu > 0 , \mu > 0, 3\beta + 2\gamma > 0 , \gamma > 0 ,$$

$$\mu + \alpha > 0 , \gamma + \varepsilon > 0 , \alpha > 0 , \varepsilon > 0 .$$

These limitations result from the fact that U is a quadratic, positively defined form. Taking into account 2.7 we obtain the following constitutive equations

$$
\begin{aligned}
\sigma_{ji} &= (\mu+\alpha)\gamma_{ji} + (\mu-\alpha)\gamma_{ij} + \lambda\delta_{ij}\gamma_{kk} , \\
\mu_{ji} &= (\gamma+\varepsilon)\varkappa_{ji} + (\gamma-\varepsilon)\varkappa_{ij} + \beta\delta_{ij}\varkappa_{kk} .
\end{aligned} \tag{2.9}
$$

Now if we eliminate the stresses from the equations of motion by means of the constitutive equations and then we make use of the defining relations for the tensors γ_{ji}, $\varkappa_{ji}$ we obtain the system of six equations in terms of the displacement $\underline{u}$ and rotation $\underline{\varphi}$.

In the vector form the equations are the following

$$
\begin{aligned}
&\Box_2 \underline{u} + (\lambda+\mu-\alpha)\,\mathrm{grad\,div}\,\underline{u} + 2\alpha\,\mathrm{curl}\,\underline{\varphi} + \underline{X} = 0 , \\
&\Box_4 \underline{\varphi} + (\beta+\gamma-\varepsilon)\,\mathrm{grad\,div}\,\underline{\varphi} + 2\alpha\,\mathrm{curl}\,\underline{u} + \underline{Y} = 0 .
\end{aligned} \tag{2.10}
$$

Here the following differential operators have been introduced

$$
\Box_2 = (\mu+\alpha)\Delta - \varrho\partial_t^2 , \qquad \Box_4 = (\gamma+\varepsilon)\Delta - 4\alpha - J\partial_t^2 .
$$

The first of these operators is d'Alembert operator, the second one Klein-Gordon's operator.

We have obtained the complex system of hyperbolic, coupled differential equations. The boundary and initial conditions should be added to the system. According to the assumption the boundary conditions have the form

$$
\begin{aligned}
&\sigma_{ji}(\underline{x},t)n_j(\underline{x}) = p_i(\underline{x},t) , \quad \mu_{ji}(\underline{x},t)n_j(\underline{x}) = m_i(\underline{x},t), \quad \underline{x}\in A_\sigma , t>0, \\
&u_i(\underline{x},t) = f_i(\underline{x},t), \quad \varphi_i(\underline{x},t) = g_i(\underline{x},t) , \quad \underline{x}\in A_u , \ t>0 .
\end{aligned} \tag{2.11}
$$

Here $\underline{n}$ is the unit vector normal to the boundary while p_i, m_i, f_i, and g_i are the given functions.
The initial conditions have the form

$$u_i(\underline{x},0) = k_i(\underline{x}),\quad \varphi_i(\underline{x},0) = t_i(\underline{x}),$$

$$\dot{u}_i(\underline{x},0) = h_i(\underline{x}),\quad \dot{\varphi}_i(\underline{x},0) = j_i(\underline{x}),\quad \underline{x}\in V,\quad t=0. \tag{2.12}$$

The coupled system of differential equations in displacements and rotations is very complicated and inconvenient to deal with, therefore our prime objective will be to uncouple it.

There are two possibilities to uncouple the equations. The first one is analogous to the method used by Lamé in the classical elastokinetics. I et us decompose the vectors $\underline{u}$ and $\underline{\varphi}$ into the potential an solenoidal parts

$$\underline{u} = \operatorname{grad}\Phi + \operatorname{curl}\underline{\Psi},\quad \operatorname{div}\underline{\Psi} = 0,$$

$$\underline{\varphi} = \operatorname{grad}\Gamma + \operatorname{curl}\underline{H},\quad \operatorname{div}\underline{H} = 0. \tag{2.13 a}$$

We apply the same procedure to the body forces and moments

$$\underline{X} = \rho(\operatorname{grad}\vartheta + \operatorname{curl}\underline{\chi}),\quad \operatorname{div}\underline{\chi} = 0,$$

$$Y = \mathfrak{J}(\operatorname{grad}\sigma + \operatorname{curl}\underline{\eta}),\quad \operatorname{div}\underline{\eta} = 0. \tag{2.13 b}$$

Substituting the above relations into Eqs. (1.10) we obtain the following simple wave equations

$$\Box_1 \Phi + \varrho \vartheta = 0 , \qquad \Box_3 \Gamma + \Im \sigma = 0$$

(2.14)
$$\Box_2 \underline{H} + 2\alpha \operatorname{curl} \underline{H} + \varrho \underline{\chi} = 0,$$

$$\Box_4 \underline{H} + 2\alpha \operatorname{curl} \underline{\Psi} + \Im \underline{\eta} = 0,$$

where we have introduced the following operators

$$\Box_1 = (\lambda + 2\mu)\Delta - \varrho \partial_t^2 , \quad \Box_3 = (\beta + 2\gamma)\Delta - 4\alpha - \Im \partial_t^2 .$$

The first of the equations represents the equation of the longitudinal wave, identical in the form to the longitudinal wave equation in the classical elastokinetics. The second equation is a new type of equation, namely the equation of the longitudinal microrotational wave. The third and fourth equations describe the propagation of the displacement shear wave and the microrotational shear wave respectively.

The longitudinal wave is well known in the classical elastokinetics. The displacement microrotational wave was investigated by N. A. Palmov [10] and W. Nowacki [14]. The last two equations of 2.14 after the elimination of $\underline{\Psi}$ and $\underline{H}$ assume the following form

(2.15)
$$(\Box_2 \Box_4 + 4\alpha^2 \Delta)\underline{\Psi} = 2\alpha \Im \operatorname{curl} \underline{\eta} - \varrho \Box_4 \underline{\chi} ,$$

$$(\Box_2 \Box_4 + 4\alpha^2 \Delta)\underline{H} = 2\alpha\varrho \operatorname{curl} \underline{\chi} - \Im \Box_2 \underline{\eta} .$$

This type of equations has been investigated by J. Ignaczak [15]. He likewise has given the "radiation conditions" similar to

Sommerfeld's conditions. It is evident that the displacement wave Γ and the shear waves $\underline{\Psi}$ and $\underline{H}$ disperse. The system of wave equations (2.14) is very useful for the determination of the singular solutions (Green functions) in an infinite space. Such solutions have been obtained, in a closed form, for the case of concentrated forces and moments harmonically varying in time by W. Nowacki and W.K. Nowacki [16]. Finally it has been shown that the assumed method of solution by means of the potentials Φ, Γ, $\underline{\Psi}$, $\underline{H}$ leads to the complete solutions (W. Nowacki [17]).

The second method of resolution of eqs. (2.10) follows that of B. Galerkin [18] in the classical elastostatics, and M. Iacovache [19] in the classical elastodynamics. The functions of this type for the dynamical problems of the micropolar elasticity have been given by N. Sandru [20], and later, in a different way, by J. Stefaniak [21]. The representation of N. Sandru has the form

$$\begin{aligned} \underline{u} &= \Box_1 \Box_4 \underline{F} - \operatorname{grad} \operatorname{div} \Xi \underline{F} - 2\alpha \operatorname{curl} \Box_3 \underline{G}, \\ \underline{\varphi} &= \Box_2 \Box_3 \underline{G} - \operatorname{grad} \operatorname{div} \Theta \underline{G} - 2\alpha \operatorname{curl} \Box_1 \underline{F}, \end{aligned} \qquad (2.16)$$

where

$$\Xi = (\lambda + \mu - \alpha) \Box_4 - 4\alpha^2, \qquad \Theta = (\beta + \gamma - \varepsilon) \Box_2 - 4\alpha^2.$$

Here the displacements $\underline{u}$ and the rotations $\underline{\varphi}$ are represented by two vector functions $\underline{F}$ and $\underline{G}$. Substituting eqs. (2.16) into eqs. (2.10) we obtain two repeated wave equations for the functions $\underline{F}$ and $\underline{G}$.

$$\begin{aligned} &\square_1 (\square_2 \square_4 + 4\alpha^2)\underline{F} + \underline{X} = 0\,, \\ &\square_3 (\square_2 \square_4 + 4\alpha^2)\underline{G} + \underline{Y} = 0\,. \end{aligned} \tag{2.17}$$

These equations are particularly useful for the determination of the displacements and the rotations generated in an infinite space by the concentrated forces and moments. So far only the singular solutions for the concentrated forces and moments varying harmonically in time have been obtained. In this case the system of equations (2.17) reduces to the system of simple elliptic equations

$$\begin{aligned} &(\Delta + \mu_1^2)(\Delta + k_1^2)(\Delta + k_2^2)\underline{F}^* + \underline{X}^* = 0\,, \\ &(\Delta + \mu_3^2)(\Delta + k_1^2)(\Delta + k_2^2)\underline{G}^* + \underline{Y}^* = 0\,, \end{aligned} \tag{2.18}$$

where $\underline{X}(\underline{x},t) = \underline{X}(\underline{x})e^{-i\omega t}$ and so on.

There exists the second way of obtaining the fundamental equations of micropolar elasticity. It consists in the utilization of the compatibility equations

$$\begin{aligned} &\gamma_{li,h} - \gamma_{hi,l} - \varepsilon_{khi}\varkappa_{lk} + \varepsilon_{kli}\varkappa_{hk} = 0\,, \\ &\varkappa_{li,h} = \varkappa_{hi,l}\,, \end{aligned} \tag{2.19}$$

and expressing the functions γ_{ji}, $\varkappa_{ji}$ by the stresses σ_{ji}, μ_{ji}.

The system of stress equations constitutes a generalization of the Beltrami-Michell equations known in the classical theory of elasticity, and has been derived for the dynamical problems by Z. Olesiak [22] and for the statical problems by N. Sandru [20]. These equations may have a practical meaning in the two-dimensional problems.

Let us consider particular cases referring to the wave propagation. Many papers have been devoted to interesting problems concerning the one-dimensional waves, dependent on x_1 and t, next dependent on $r=(x_1^2+x_2^2)^{\frac{1}{2}}$ and t, and dependent on $R=(x_1^2+x_2^2+x_3^2)^{\frac{1}{2}}$ and t. Here we should mention the papers by A. C. Eringen [23] N. A. Palmov [10] and A. C. Smith [24].

Consider two-dimensional problems. Let us assume that we have to deal with the problem for which the displacements are independent of x_3. In such a case the system of equations (2.10) can be decomposed into two mutually independent systems of equations. In the first system of equations the following vectors occur

$$\underline{u} = (u_1, u_2, 0), \quad \underline{\varphi} = (0,0,\varphi_3), \quad \underline{X} = (X_1, X_2, 0), \quad \underline{Y} = (0,0,Y_3) \qquad (2.20)$$

Now the system of equations takes the form

$$\begin{aligned} &(\mu+\alpha)\nabla_1^2 u_1 - \varrho\ddot{u}_1 + (\mu+\lambda-\alpha)\partial_1 e + 2\alpha\partial_2\varphi_3 + X_1 = 0,\\ &(\mu+\alpha)\nabla_1^2 u_2 - \varrho\ddot{u}_2 + (\mu+\lambda-\alpha)\partial_2 e - 2\alpha\partial_1\varphi_3 + X_2 = 0,\\ &\left[(\gamma+\varepsilon)\nabla_1^2 - 4\alpha - \mathfrak{J}\partial_t^2\right]\varphi_3 + 2\alpha(\partial_1 u_2 - \partial_2 u_1) + Y_3 = 0, \end{aligned} \qquad (2.21)$$

where $\nabla_1^2 = \partial_1^2 + \partial_2^2$, $\quad e = \partial_1 u_1 + \partial_2 u_2$.

The field of displacements $(u_1, u_2, 0)$ and rotations $(0,0,\varphi_3)$ generates in the body the state of stresses described by the following matrices

$$(2.22)\qquad \underline{\sigma} = \begin{Vmatrix} \sigma_{11}, & \sigma_{21}, & 0 \\ \sigma_{21}, & \sigma_{22}, & 0 \\ 0, & 0, & \sigma_{33} \end{Vmatrix}, \quad \underline{\mu} = \begin{Vmatrix} 0, & 0, & \mu_{13} \\ 0, & 0, & \mu_{23} \\ \mu_{31}, & \mu_{32}, & 0 \end{Vmatrix} .$$

In the second system, determined by the vectors

$$(2.23)\qquad \underline{u} = (0,0,u_3),\ \underline{\varphi} = (\varphi_1,\varphi_2,0),\ \underline{X} = (0,0,X_3),\ Y = (Y_1,Y_2,0)$$

we have to deal with the system of equations

$$(2.24)\qquad \begin{aligned} &[(\gamma+\varepsilon)\nabla_1^2 - 4\alpha - J\partial_t^2]\varphi_1 + (\gamma+\beta-\varepsilon)\partial_1\varkappa + 2\alpha\partial_2 u_3 + Y_1 = 0, \\ &[(\gamma+\varepsilon)\nabla_1^2 - 4\alpha - J\partial_t^2]\varphi_2 + (\gamma+\beta-\varepsilon)\partial_2\varkappa - 2\alpha\partial_1 u_3 + Y_2 = 0, \\ &(\mu+\alpha)\nabla_1^2 u_3 - \rho\ddot{u}_3 + 2\alpha(\partial_1\varphi_2 - \partial_2\varphi_1) + X_3 = 0, \end{aligned}$$

where

$$\varkappa = \partial_1\varphi_1 + \partial_2\varphi_2 .$$

It is easy to verify that the matrices

$$(2.25)\qquad \underline{\sigma} = \begin{Vmatrix} 0, & 0, & \sigma_{13} \\ 0, & 0, & \sigma_{23} \\ \sigma_{31}, & \sigma_{32}, & 0 \end{Vmatrix}, \quad \underline{\mu} = \begin{Vmatrix} \mu_{11}, & \mu_{12}, & 0 \\ \mu_{21}, & \mu_{22}, & 0 \\ 0, & 0, & \mu_{23} \end{Vmatrix},$$

correspond to the field of displacement $\underline{u} = (0,0,u_3)$ and rotation $\underline{\varphi} = (\varphi_1, \varphi_2, 0)$.

Let us dwell our attention on the first system of equations. Introducing the potentials Φ and Ψ , where

$$u_1 = \partial_1\Phi - \partial_2\Psi \ , \quad u_2 = \partial_2\Phi - \partial_1\Psi \ , \ \varphi_3 = \varphi \ , \tag{2.26}$$

the system of equations 2.21 is reduced to simple wave equations (for $X = Y = 0$)

$$[(\lambda + 2\mu)\nabla_1^2 - \rho\partial_t^2]\Phi = 0 \ ,$$
$$\{[(\mu+\alpha)\nabla_1^2 - \rho\partial_t^2][(\gamma+\varepsilon)\nabla_1^2 - 4\alpha - J\partial_t^2] + 4\alpha^2\nabla^2\}(\Psi, \varphi) = 0 \ . \tag{2.27}$$

Many authors have investigated the above system of equations. V. R. Parfitt and A. C. Eringen [25] and J. Stefaniak [26] have investigated the reflection of a plane wave from the free boundary of an infinite space. A. C. Eringen and E. S. Suhubi [11] investigated the Rayleigh wave, generalized in the micropolar continuum. The same problem is discussed in the extensive paper by S. Kaliski, J. Kupelewski and C. Rymarz [27]. The wave propagation in a plate (the generalized Lamb's problem) has been considered by W. Nowacki and W. K. Nowacki [28]. Also a number of boundary value problems have been solved for the case when the loadings harmonically varying in time act on the boundary of an elastic semi-space (W. Nowacki and W. K. Nowacki [29], G. Eason [30]). Finally we notice the trends to solve the approximate wave equations (G. Eason [31], J. D. Achenbach [32]) and the interesting results obtained in this way.

Let us return to the second system of two-dimensional equations for which the deformation is determined by the vectors $\underline{u} = (0, 0, u_3)$, $\varphi = (\varphi_1, \varphi_2, 0)$. By means of the potentials Γ , H the system of equations (2.24) is reduced to simple wave equations

(2.28) $$\varphi_1 = \partial_1\Gamma - \partial_2 H\ , \quad \varphi_2 = \partial_2\Gamma + \partial_1 H\ ,$$

The equations take the following form

(2.29)
$$[(\beta+2\gamma)\nabla_1^2 - 4\alpha - J\partial_t^2]\Gamma = 0\ ,$$
$$\{[(\mu+\alpha)\nabla_1^2 - \rho\partial_t^2][(\gamma+\varepsilon)\nabla_1^2 - 4\alpha - J\partial_t^2] + 4\alpha^2\nabla_1^2\}(H, u_3) = 0\ .$$

The first equation corresponds to the longitudinal microrotational wave, the second one to the shear wave. If we assume that

(2.30) $$(\varphi_1, \varphi_2, u_3) = (\varphi_1^*(x_1), \varphi_2^*(x_1), u_3^*(x_1))e^{i(kx_2-\omega t)}\ ,$$

and the boundary of the elastic semi-space $x_1 = 0$ is free from stresses, the above functions lead to the Love surface waves. The propagation of these waves have been investigated in the paper [27] . It is interesting to note that within the frames of the classical elasticity Love's waves do not exist in the case of the homogeneous elastic semi space, the propagation of such waves is possible only for a layered semi-space and different densities and Lame's constants of both media.

Let us consider the second type of the two-dimensional problems, namely the problems of the axially symmetric deformations. In this case the system of equations (2.10) can be decomposed into two mutually independent systems of equations. The following vectors enter the first system of equations

$$\underline{u} = (u_r, 0, u_z), \ \underline{\varphi} = (0, \varphi_\theta, 0), \ \underline{X} = (X_r, 0, X_z), \ \underline{Y} = (0, Y_\theta, \theta) \quad (2.31)$$

The system of equations takes the form

$$\begin{aligned} &\left[(\mu+\alpha)\left(\nabla^2 - \frac{1}{r^2}\right)u_r - \rho\ddot{u}_r\right] + (\lambda+\mu-\alpha)\frac{\partial e}{\partial r} - 2\alpha\frac{\partial \varphi_\theta}{\partial r} + X_r = 0, \\ &[(\mu+\alpha)\nabla^2 - \rho\partial_t^2]u_z + (\lambda+\mu-\alpha)\frac{\partial e}{\partial z} + 2\alpha\frac{1}{r}\frac{\partial}{\partial r}(r\varphi_\theta) + X_z = 0, \quad (2.32) \\ &\left[(\gamma+\varepsilon)\left(\nabla^2 - \frac{1}{r^2}\right) - 4\alpha - J\partial_t^2\right]\varphi_\theta + 2\alpha\left(\frac{\partial u_r}{\partial z} - \frac{\partial u_z}{\partial r}\right) + Y_\theta = 0, \end{aligned}$$

here

$$\nabla^2 = \frac{\partial^2}{\partial r^2} + \frac{1}{r}\frac{\partial}{\partial r} + \frac{\partial^2}{\partial z^2}, \quad e = \frac{1}{r}\frac{\partial}{\partial r}(r u_r) + \frac{\partial u_z}{\partial z}.$$

The following stress matrices correspond to the deformation presented here

$$\underline{\sigma} = \begin{Vmatrix} \sigma_{rr}, & 0, & \sigma_{rz} \\ 0, & \sigma_{\theta\theta}, & 0 \\ \sigma_{zr}, & 0, & \sigma_{zz} \end{Vmatrix}, \quad \underline{\mu} = \begin{Vmatrix} 0, & \mu_{r\theta}, & 0 \\ \mu_{\theta r} & 0, & \mu_{\theta z} \\ 0, & \mu_{z\theta}, & 0 \end{Vmatrix}. \quad (2.33)$$

The following vectors occur in the second system of equations

$$\underline{u} = (0, u_\theta, 0), \ \varphi = (\varphi_r, 0, \varphi_z), \ \underline{X} = (0, X_\theta, 0), \ \underline{Y} = (Y_r, 0, Y_z), \quad (2.34)$$

and the stress matrices

$$(2.35)\qquad \sigma = \begin{Vmatrix} 0 & \sigma_{r\theta} & 0 \\ \sigma_{\theta r} & 0 & \sigma_{\theta z} \\ 0 & \sigma_{z\theta} & 0 \end{Vmatrix}, \quad \mu = \begin{Vmatrix} \mu_{rr} & 0 & \mu_{rz} \\ 0 & \mu_{\theta\theta} & 0 \\ \mu_{zr} & 0 & \mu_{zz} \end{Vmatrix}$$

Now the system of equations assumes the form

$$\Big[(\gamma+\varepsilon)\Big(\nabla^2-\frac{1}{r^2}\Big)-4\alpha-J\partial_t^2\Big]\varphi_r+(\beta+\gamma-\varepsilon)\frac{\partial\varkappa}{\partial r}-2\alpha\frac{\partial u_\theta}{\partial z}+Y_r=0,$$

$$(2.36)\qquad [(\gamma+\varepsilon)\nabla^2-4\alpha-J\partial_t^2]\varphi_z+(\beta+\gamma-\varepsilon)\frac{\partial\varkappa}{\partial z}+\frac{2\alpha}{r}\frac{\partial}{\partial r}(ru_\theta)+Y_z=0,$$

$$\Big[(\mu+\alpha)\Big(\nabla^2-\frac{1}{r^2}\Big)-\rho\partial_t^2\Big]u_\theta+2\alpha\Big(\frac{\partial\varphi_r}{\partial z}-\frac{\partial\varphi_z}{\partial r}\Big)+X_\theta=0,$$

where

$$\varkappa=\frac{1}{r}\partial/\partial r\,(r\varphi_r)+\partial\varphi_z/\partial z\,.$$

Both the systems of equations can be reduced to simple wave equations. These equations served in the investigation of longitudinal and torsional waves in an infinite cylinder of circular cross-section, and in solving two generalized axially symmetric Lamb's problems (W. Nowacki and W. K. Nowacki [33], [34]).

Concluding this review of the dynamical problems we should mention the general theorems of the micropolar elastokinetics. These theorems have been presented and derived by a number of authors.

The theorem on the reciprocity of work has the form (N. Sandru [20], D. Iesan [35])

$$\int_V (X_i * u_i' + Y_i * \varphi_i')\,dV + \int_A (p_i * u_i' + m_i * \varphi_i')\,dA =$$
$$\int_V (X_i' * u_i + Y_i' * \varphi_i)\,dV + \int_A (p_i' * u_i + m_i' * \varphi_i)\,dA, \tag{2.37}$$

where

$X_i * u_i' = \int_0^t X_i(\underline{x}, t-\tau)\,u_i'(\underline{x},\tau)\,d\tau$ and so on are the convolutions. This equation constitutes a generalization of Graffi's theorem [36] known in the classical elastokinetics. The theorem on the reciprocity constitutes one of the most interesting theorems in the micropolar theory of elasticity. The theorem is extremely general and includes the possibility of derivating the method of integration of the equation of elastokinematics by means of Green's function.

The principle of virtual work is of considerable importance

$$\int_V [(X_i - \varrho\ddot{u})\delta u_i + (Y_i - J\ddot{\varphi}_i)\delta\varphi_i]\,dV +$$
$$+ \int_A (p_i \delta u_i + m_i \delta\varphi_i)\,dA = \int_V (\sigma_{ji}\delta\gamma_{ji} + \mu_{ji}\delta\varkappa_{ji})\,dV. \tag{2.38}$$

Here δu_i and $\delta\varphi_i$ denote the virtual displacements and rotations. The principle of virtual work may serve for the derivation of the equation of plate and shell bending under the suitable approximations, for the approximate solution of the equations of elastokinetics and finally for the derivation of the uniqueness theorem.

An important role is played by the extended

Hamilton's principle

$$\delta\int_{t_1}^{t_2}(W - \mathcal{K})dt = \int_{t_1}^{t_2}\delta\mathcal{L}\,. \tag{2.39}$$

Here we assume that $\delta\underline{u}(\underline{x},t_1) = \delta\underline{u}(\underline{x},t_2) = \delta\underline{\varphi}(\underline{x},t_1) = \delta\underline{\varphi}(\underline{x},t_2) = 0.$

$\delta\mathcal{L}$ denotes the virtual work of external forces

$$\delta\mathcal{L} = \int_V (X_i\delta u_i + Y_i\delta\varphi_i)dV + \int_A (p_i\delta u_i + m_i\delta\varphi_i)dA\,,$$

and $\mathcal{K}$ is the kinetic energy. W is the work of deformation which, in our case of the adiabatic process, is identical to the internal energy $\mathcal{U}$.

In the present review we have only shown the most important, in our opinion, results of the micropolar elasticity. Let us note that the fundamental results have been obtained only in the case of propagation of the monochromatic waves. The investigation concerning the problems of the waves generated by aperiodic causes or by the causes moving with constant or varying speed have been hardly initiated. The contemporary investigations of the dynamical problems tend to include also the other physical fields. The research in the domain of micropolar thermoelasticity and micropolar magnetoelasticity is already developing.

3. The Micropolar Elastostatics

The substitution of the constitutive equations into the equilibrium equations, together with the definition of the deformations taken into account, leads to the system of six differential equations in terms of displacements and rotations. In the compact vector form the equations read

$$
\begin{gathered}
(\mu+\alpha)\nabla^2\underline{u} + (\lambda+\mu-\alpha)\,\mathrm{grad\,div}\,\underline{u} + 2\alpha\,\mathrm{curl}\,\underline{\varphi} + \underline{X} = 0, \\
[(\gamma+\varepsilon)\nabla^2 - 4\alpha]\underline{\varphi} + (\beta+\gamma-\varepsilon)\,\mathrm{grad\,div}\,\underline{\varphi} + 2\alpha\,\mathrm{curl}\,\underline{u} + \underline{Y} = 0.
\end{gathered}
\tag{3.1}
$$

The system is coupled, of elliptic type. Let us note that the material constants μ, λ, α, β, γ, ε, occurring in the equations refer to the isothermal process. The system of equations can be decomposed into two independent systems of equations only in the particular case $\alpha = 0$. We can put the following question: is it possible to compose the solution to the system of equations of two parts, the first of which has exactly the same form as the solution of the classical elastostatics. An affirmative answer to this question has been given by H. Schaefer [37].

Introducing the vector

$$
\underline{\zeta} = \frac{1}{2}\,\mathrm{curl}\,\underline{u} - \underline{\varphi}
\tag{3.2}
$$

and eliminating the function $\underline{\varphi}$ from the system of homogeneous equations (3.1) we obtain

$$\mu\nabla^2\underline{u} + (\lambda+\mu)\,\mathrm{grad\,div}\,\underline{u} = 2\alpha\,\mathrm{curl}\,\underline{\zeta}\,,$$

(3.3)

$$[(\gamma+\varepsilon)\nabla^2 - 4\alpha]\underline{\zeta} + (\beta+\gamma-\varepsilon)\,\mathrm{grad\,div}\,\underline{\zeta} = \frac{1}{2}(\gamma+\varepsilon)\nabla^2\mathrm{curl}\,\underline{u}\,.$$

We assume the solution to this system in the form

(3.4) $$\underline{u} = \underline{u}' + \underline{u}''\,,\quad \underline{\zeta} = \underline{\zeta}' + \underline{\zeta}''\,,\qquad \text{where}\qquad \underline{\zeta}' = 0\,.$$

The above representation allows us to split the system of equations (3.3) into two independent systems of equations

(3.5) $$\mu\nabla^2\underline{u}' + (\lambda+\mu)\,\mathrm{grad\,div}\,\underline{u}' = 0\,,\qquad \nabla^2\mathrm{curl}\,\underline{u}' = 0\,,$$

and

$$\mu\nabla^2\underline{u}'' + (\lambda+\mu)\,\mathrm{grad\,div}\,\underline{u}'' = 2\alpha\,\mathrm{curl}\,\underline{\zeta}''\,,$$

(3.6)

$$[(\gamma+\varepsilon)\nabla^2 - 4\alpha]\underline{\zeta}'' + (\beta+\gamma-\varepsilon)\,\mathrm{grad\,div}\,\underline{\zeta}'' = \frac{1}{2}(\gamma+\varepsilon)\nabla^2\mathrm{curl}\,\underline{u}''.$$

Let us point out that the system of equations (3.5) in its form is identical with the corresponding system of equations of the classical theory of elasticity.

Let us assume that on the boundary of the body the loadings $\underline{p}$ and moments $\underline{m}$ are prescribed

(3.7) $$p_i = \sigma_{ji}n_j\,,\qquad m_i = \mu_{ji}n_j\,.$$

The system of equations (3.6) is satisfied with the boundary conditions $p_i = \sigma'_{ji}n_j$. The assumption $\underline{\zeta}' = 0$ is synonimous with the assertion that the skew-symmetric part of the tensor

γ'_{ji} is equal to zero $(\gamma_{\langle ij \rangle} = 0)$. The tensor γ'_{ji} is thus symmetric. Therefore the strains σ'_{ji} compose the symmetric tensor. However the assumption $\underline{\zeta}' = 0$ leads to the relation $\underline{\varphi}' = \frac{1}{2} \operatorname{curl} \underline{u}'$. Since $\underline{\varphi}' \neq 0$ therefore also $\varkappa'_{ji} \neq 0$. Hence the following moment stress exist in the body

$$\mu'_{ji} = 2\gamma \varkappa_{(ji)} + 2\varepsilon \varkappa'_{\langle ji \rangle} + \beta \varkappa'_{kk} \delta_{ij} . \tag{3.8}$$

As a rule the condition $m_i = \mu'_{ji} n_j$ does not hold. Since the functions $\underline{u}'$ do not satisfy all the boundary conditions the solutions $\underline{u}''$, $\underline{\zeta}''$ satisfying the system of equations (3.6) and the boundary conditions

$$\sigma''_{ji} n_j = 0 \quad , \quad (\mu_{ji} + \mu''_{ji}) n_j = m_i , \tag{3.9}$$

should be added to the solutions $\underline{u}'$, $\underline{\zeta}'$.

H. Schaefer assumed the solution of eqs. (3.6) in the form

$$\underline{\zeta}'' = \operatorname{grad} \Phi + \operatorname{curl} \operatorname{curl} \underline{\Omega} \tag{3.10}$$

where the functions Φ, $\underline{\Omega}$ satisfy the differential equations

$$(h^2 \nabla^2 - 1)\Phi = 0 , \quad (\nu^2 \nabla^2 - 1)\underline{\Omega} = 0 ,$$
$$h^2 = \frac{2\gamma + \beta}{4\alpha} , \quad \nu^2 = \frac{\gamma + \varepsilon}{4\alpha} . \tag{3.11}$$

However the proof of the completeness of the solutions $\underline{\zeta}''$ is lacking. Here, as in the elastokinetics, the displacements and rotations can be represented by two vector functions $\underline{G}$, $\underline{F}$ constituting a generalization of Galerkin's functions. If we substitute the representation given by N. Sandru [20]

$$
(3.12)\quad
\begin{aligned}
\underline{u} &= (\lambda+2\mu)\nabla^2[(\gamma+\varepsilon)\nabla^2-4\alpha]\underline{F}-[(\gamma+\varepsilon)(\lambda+\mu-\alpha)\nabla^2+\\
&\quad -4\alpha(\lambda+\mu)]\,\mathrm{grad\,div}\,\underline{F}-2\alpha[(\beta+2\gamma)\nabla^2-4\alpha]\,\mathrm{curl}\,\underline{G}\ ,\\
\underline{\varphi} &= (\mu+\alpha)\nabla^2[(\beta+2\gamma)\nabla^2-4\alpha]\underline{G}-[(\mu+\alpha)(\beta+\gamma-\varepsilon)\nabla^2+\\
&\quad -4\alpha]\,\mathrm{grad\,div}\,\underline{G}-2\alpha(\lambda+2\mu)\nabla^2\,\mathrm{curl}\,\underline{F}\ ,
\end{aligned}
$$

into the system of equations (3.1) we obtain the following simple equations for the functions $\underline{F}$ and $\underline{G}$

$$
(3.13)\quad
\begin{aligned}
&(\lambda+2\mu)\nabla^2\nabla^2(\ell^2\nabla^2-1)\underline{F}+\underline{X}=0\ ,\\
&16\alpha^2\mu\nabla^2(h^2\nabla^2-1)(\ell^2\nabla^2-1)\underline{G}+\underline{Y}=0\ .
\end{aligned}
$$

Let us observe that the assumption $\underline{X}=0$ entails also that $\underline{F}=0$. Similarly for $\underline{Y}=0$ we have $\underline{G}=0$. Eqs. (3.13) allow us to determine, in a very simple way, the Green functions in an infinite micropolar space. Below we list only the final results of the singular solutions. Let a concentrated unit force, directed along the x_i axis, act at a point $\underline{\xi}$. The displacements and rotations generated by the force take the following form [20]

$$
(3.14)\quad
\begin{aligned}
u_j^{(i)} &= -\frac{\lambda+\mu}{8\pi\mu(\lambda+2\mu)}\partial_i\partial_j\left[R\frac{(\gamma+\varepsilon)(\lambda+2\mu)}{2\mu(\lambda+\mu)}\left(\frac{1-e^{-\frac{R}{\ell}}}{R}\right)\right]-\frac{1}{4\mu\pi}\left(\frac{\alpha}{\mu+\alpha}\frac{e^{-\frac{R}{\ell}}}{R}-\frac{1}{R}\right)\delta_{ij}\\
\varphi_j^{(i)} &= \frac{1}{8\pi\mu}\epsilon_{ijk}\frac{\partial}{\partial x_k}\left(\frac{1-e^{-\frac{R}{\ell}}}{R}\right).
\end{aligned}
$$

Here R is the distance between the points $\mathbf{x}$ and $\boldsymbol{\xi}$. Passing to the classical theory of elasticity we have

$$u_j^{(i)} = -\frac{\lambda+\mu}{8\pi\mu(\lambda+2\mu)}\partial_i\partial_j(R) + \frac{1}{4\pi\mu R}\delta_{ij}\,, \quad \varphi_j^{(i)} = 0\,.$$

In the case of a concentrated unit moment applied at a point $\boldsymbol{\xi}$ and acting in the direction of the x_i axis we have the following displacements $\hat{u}_j^{(i)}$ and rotations $\hat{\varphi}_j^{(i)}$ [20] :

$$\hat{u}_j^{(i)} = \frac{1}{8\pi\mu}\epsilon_{ijk}\frac{\partial}{\partial x_k}\left(\frac{1-e^{-\frac{R}{\ell}}}{R}\right)\,,$$
$$\hat{\varphi}_j^{(i)} = \frac{1}{16\pi\mu}\partial_i\partial_j\left[\frac{1-e^{-\frac{R}{\ell}}}{R} + \frac{\mu}{\alpha}\left(\frac{e^{-\frac{R}{h}} - e^{-\frac{R}{\ell}}}{R}\right)\right] + \frac{e^{-\frac{R}{\ell}}}{4\pi(\gamma+\varepsilon)R}\delta_{ij}\,. \tag{3.15}$$

Beside the Galerkin function also the Papkovitch-Neuber type functions have been introduced in the micropolar elastostatics. H. Neuber [8] has generalized his functions on micropolar elastostatics and applied them to a series of problems referring to the stress concentration problems around holes and notches [38] - [40]. A certain variation of this type function has been given by N. Sandru [20] and S. C. Cowin [41] .

Parallely to the equations in terms of the displacements and rotations in elastostatics one can use the stress equations analogous to those of Beltrami-Mitchell. In this respect the discussion of H. Schaefer is interesting, he introduced a very general type of stress functions known in the classical elastostatics [42]. Also the paper of S. Kassel [51] on the stress functions deserves attention.

In a more detailed way we shall discuss the two-dimensional problems, namely the problems of the plane state of strain, and the axially symmetric problems. Consider the plane state of strain for which the deformation does not depend on the variable x_3. As we know, in this case two mutually independent systems of equations are obtained. In the first problem there appear the vectors $\underline{u} = (u_1, u_2, 0)$ $\underline{\varphi} = (0, 0, \varphi_3)$, while in the second one the vectors $\underline{u} = (0, 0, u_3)$, $\underline{\varphi} = (\varphi_1, \varphi_2, 0)$.

The following compatibility equations

$$
\begin{aligned}
&\partial_1^2 \sigma_{22} + \partial_2^2 \sigma_{11} - \frac{\lambda}{2(\lambda+\mu)} \nabla^2(\sigma_{11} + \sigma_{22}) = \partial_1 \partial_2 (\sigma_{12} + \sigma_{21}) \, , \\
&(\partial_2^2 - \partial_1^2)(\sigma_{12} + \sigma_{21}) + \frac{\mu}{\alpha} \nabla^2 (\sigma_{12} - \sigma_{21}) + 2\partial_1 \partial (\sigma_{11} - \sigma_{22}) + \\
&+ 4\mu/(\gamma + \varepsilon) \cdot (\partial_1 \mu_{13} + \partial_2 \mu_{23}) = 0 \, , \quad \partial_1 \mu_{23} = \partial_2 \mu_{13} \, ,
\end{aligned}
\tag{3.16}
$$

and the equilibrium equations

$$
\begin{aligned}
&\partial_1 \sigma_{11} + \partial_2 \sigma_{21} = 0 \, , \qquad \partial_1 \sigma_{12} + \partial_2 \sigma_{22} = 0 \, , \\
&\sigma_{12} - \sigma_{21} + \partial_1 \mu_{13} + \partial_2 \mu_{23} = 0 \, . \, ,
\end{aligned}
\tag{3.17}
$$

constitute the point of departure for the first problem. We have the system of six equations for the determination of six unknown functions σ_{11}, σ_{22}, σ_{12}, σ_{21}, μ_{13}, μ_{23}. The equilibrium equations are satisfied identically by two functions F and Ψ connected with the stresses by the following relations

$$\sigma_{11} = \partial_2^2 F - \partial_1\partial_2 \Psi \,, \qquad \sigma_{22} = \partial_1^2 F + \partial_1\partial_2 \Psi \,,$$

$$\sigma_{12} = -\partial_1\partial_2 F - \partial_2^2 \Psi \,, \qquad \sigma_{21} = -\partial_1\partial_2 F + \partial_1^2 \Psi \,, \tag{3.18}$$

$$\mu_{\alpha 3} = \partial_\alpha \Psi \,, \qquad \alpha, \beta = 1, 2$$

The function F is Airy's function known from the classical elastokinetics. The function Ψ has been introduced by R.D. Mindlin [43] and H. Schaefer [7] for the plane state of strain in the Cosserat continuum and pseudo-continuum. Substituting eqs.(3.18) into the compatibility equations we obtain the following equations in terms of the functions Ψ and F

$$\nabla_1^2 \nabla_1^2 F = 0 \quad , \qquad \nabla_1^2 (l^2 \nabla_1^2 - 1)\Psi = 0 \,. \tag{3.19}$$

The functions F and Ψ are mutually dependent and satisfy the Cauchy-Riemann conditions

$$\begin{aligned} \partial_1 (l^2 \nabla_1^2 - 1)\Psi &= A \partial_2 \nabla_1^2 F \,, \\ \partial_2 (l^2 \nabla_1^2 - 1)\Psi &= -A \partial_1 \nabla_1^2 F \,. \end{aligned} \qquad A = \frac{(\lambda + 2\mu)(\gamma + \varepsilon)}{4\mu(\lambda + \mu)} \,. \tag{3.20}$$

Since the functions $\nabla_1^2 F$ and $(l^2 \nabla_1^2 - 1)\Psi$ are harmonic, it is not difficult to observe that the method of complex variable is particularly useful for solving eqs.(3.19). This method has been successfully applied by G.N. Savin [44]- [46] and his co-workers in the problems of stress concentration around the holes. D.E. Carlson [47] has investigated the completeness of the solutions by means of the functions F and Ψ . Let us

Observe that the form of the plane problem is identical for both the pseudo-continuum and the continuum of Cosserats'. This is why there exists a number of special problems solved concerning the stress concentration, the state of stress in an elastic semi-space, and the singular solutions. First of all we mention here the papers by R. D. Mindlin [48], H. Schaefer [7], R. Muki and E. Sternberg [49], P. N. Kaloni and T. Ariman [50].

The first plane problem can be solved also by another method suitable in the case of the displacements and rotations prescribed on the boundary [52]. Using the differential equations in terms of displacements and rotations

$$(\mu+\alpha)\nabla_1^2 u_1 + (\lambda+\mu-\alpha)\partial_1 e + 2\alpha\,\partial_2\varphi_3 = 0\,,$$

$$(3.21)\qquad (\mu+\alpha)\nabla_1^2 u_2 + (\lambda+\mu-\alpha)\partial_2 e - 2\alpha\,\partial_1\varphi_3 = 0\,,$$

$$[(\gamma+\varepsilon)\nabla_1^2 - 4\alpha]\varphi_3 + 2\alpha(\partial_1 u_2 - \partial_2 u_1) = 0\,,\quad e = \partial_1 u_1 + \partial_2 u_2\,,$$

and introducing the potentials Φ, Ψ related to the displacements

$$(3.22)\qquad u_1 = \partial_1\Phi + \partial_2\Psi\,,\qquad u_2 = \partial_2\Phi - \partial_1\Psi$$

we obtain the following simple differential equations

$$(3.23)\qquad \nabla_1^2\nabla_1^2\Phi = 0\,,\qquad \nabla_1^2(l^2\nabla_1^2 - 1)\varphi_3 = 0\,.$$

The functions Φ and φ_3 are not mutually independent, they should satisfy the conditions in the form

$$\partial_1 \nabla_1^2 \Phi = -\frac{2\mu}{\lambda+2\mu}\partial_2(l^2\nabla_1^2 - 1)\varphi_3 ,$$
$$\partial_2 \nabla_1^2 \Phi = \frac{2\mu}{\lambda+2\mu}\partial_1(l^2\nabla_1^2 - 1)\varphi_3 . \qquad (3.24)$$

The conditions (3.24) are the Cauchy-Riemann conditions for the functions $\nabla_1^2\Phi$ and $(l^2\nabla_1^2 - 1)\varphi_3$.

The potential Ψ is related to the function φ_3 :

$$\nabla_1^2 \Psi = \frac{1}{2\alpha}[(\gamma+\varepsilon)\nabla^2 - 4\alpha]\varphi_3 . \qquad (3.25)$$

The solution of the first plane problem is obtained by the following procedure. Solving eqs. (3.23) for example for an elastic semi-space we get four constants of integration. To determine the constants we have three boundary conditions and the Cauchy-Riemann conditions.

In the case of the second plane state of strain we have the following system of equations

$$(\mu+\alpha)\nabla_1^2 u_3 + 2\alpha(\partial_1\varphi_2 - \partial_2\varphi_1) = 0 ,$$
$$[(\gamma+\varepsilon)\nabla_1^2 - 4\alpha]\varphi_1 + (\beta+\gamma-\varepsilon)\partial_1 x + 2\alpha\partial_2 u_3 = 0 , \qquad (3.26)$$
$$[(\gamma+\varepsilon)\nabla_1^2 - 4\alpha]\varphi_2 + (\beta+\gamma-\varepsilon)\partial_2 x + 2\alpha\partial_1 u_3 = 0 .$$

The simplest way to solve the above system of equations is to make use of the potentials Ω , Ξ [53] , where

(3.27) $$\varphi_1 = \partial_1 \Omega + \partial_2 \Xi \ , \quad \varphi_2 = \partial_2 \Omega - \partial_1 \Xi \ .$$

Substituting the relations (3.21) intc eqs. (3.20) we cbtain the system of equations

(3.28) $$\nabla_1^2 (h^2 \nabla_1^2 - 1) \Omega = 0 \ , \quad \nabla_1^2 (l^2 \nabla_1^2 - 1) \Xi = 0 \ .$$

The functions Ω and Ξ satisfy the following Cauchy-Riemann conditions

(3.29) $$\begin{aligned} -\partial_1 (h^2 \nabla_1^2 - 1) \Omega &= \frac{\mu}{\mu + \alpha} \partial_2 (l^2 \nabla_1^2 - 1) \Xi \ , \\ \partial_2 (h^2 \nabla_1^2 - 1) \Omega &= \frac{\mu}{\mu + \alpha} \partial_1 (l^2 \nabla_1^2 - 1) \Xi \ . \end{aligned}$$

The quantity u_3 is related with the potential Ξ by the equation

(3.30) $$\nabla_1^2 u_3 = \frac{2\alpha}{\mu + \alpha} \nabla_1^2 \Xi \ .$$

Another method of solution of the "second" problem of the plane state of strain belongs to M. Suchar [54]. Five compatibility equations and three equilibrium equations serve as the point of departure of his discussion. The system of equilibrium equations is satisfied by four functions Φ, Ψ, χ, Ω related to the stresses by the following formulae

(3.31) $$\begin{aligned} &\sigma_{13} = \partial_2 \Phi \ , \quad \sigma_{23} = -\partial_1 \Phi \ , \quad \sigma_{31} = -\partial_2 \Phi + \partial_1 \Psi \ , \quad \sigma_{32} = \partial_1 \Phi + \partial_2 \Psi \ , \\ &\mu_{11} = 2\Phi + \partial_2 \chi \ , \quad \mu_{21} = \Psi - \partial_1 \chi \ , \quad \mu_{12} = -\Psi + \partial_2 \Omega \ , \quad \mu_{22} = 2\Phi - \partial_1 \Omega \ . \end{aligned}$$

Substituting these relations into the equations of compatibility one obtains the following differential equations in terms of the functions Φ , Ψ , χ , and Ω

$$\nabla_1^2(h^2\nabla_1^2-1)\Phi=0\,,\quad \nabla_1^2(l^2\nabla_1^2-1)\Psi=0\,,\quad L(\chi)=0\,,\quad L(\Omega)=0, \tag{2.32}$$

where

$$L(\ )=(h^2\nabla^2-1)(l^2\nabla^2-1)\nabla^2\nabla_1^2(\ )\,.$$

One should add that the stress functions Φ , Ψ , χ , Ω are not mutually independent but combined by four additional differential relations.

In turn we consider the axially symmetric problems. We know from the preceding point that the system of six differential equations in terms of displacement and rotations can be split, in this case, into two mutually independent systems of equations. In the first one the deformation is determined by the vectors $\underline{u}=(u_r, 0, u_z)$, $\underline{\varphi}=(0, \varphi_\theta, 0)$ while in the second is given by $\underline{u}=(0, u_\theta, 0)$, $\underline{\varphi}=(\varphi_r, 0, \varphi_z)$.

In the classical theory of elasticity, solving the first problem we are very frequently making use of the Love function $\chi(r,z)$ satisfying the biharmonic equation. In the micropolar theory of elasticity we introduce two functions of Love's type. In the first axially symmetric problem the function $\chi(r,z)$ is related to the displacements and rotations in the following manner [55]

$$u_r = -\frac{\partial^2}{\partial r \partial z}(\Gamma \chi) \quad , \quad \varphi_\theta = 2\alpha(\lambda + 2\mu)\frac{\partial}{\partial r}\nabla^2 \chi \quad ,$$

(3.33)

$$u_z = \Theta \chi - \frac{\partial}{\partial z^2}(\Gamma \chi) \quad .$$

Here

$$\Gamma = (\gamma + \varepsilon)(\lambda + \mu - \alpha)\nabla^2 - 4\alpha(\lambda + \mu) \, , \; \Theta = (\lambda + 2\mu)\nabla^2[(\gamma + \varepsilon)\nabla^2 - 4\alpha] \, ,$$

$$\nabla^2 = \frac{\partial^2}{\partial r^2} + \frac{1}{r}\frac{\partial}{\partial r} + \frac{\partial^2}{\partial z^2} \, , \quad r = (x_1^2 + x_2^2)^{1/2} \, .$$

Susbtituting the above relations intc the equations of the elastostatics we obtain the following equations in terms of the function χ

(3.34) $$4\alpha\mu(\lambda + 2\mu)\nabla^2\nabla^2(l^2\nabla^2 - 1)\chi(r, z) + X_z(r, z) = 0 \, .$$

Similarly, in the second axially symmetric problem we assume

$$\varphi_r = -\frac{\partial^2}{\partial r \partial z}(\Omega \Psi) \quad , \quad u_\theta = 8\alpha^2 \frac{\partial}{\partial r}(h^2\nabla^2 - 1)\Psi \quad ,$$

(3.35)

$$\varphi_z = \Phi \Psi - \frac{\partial}{\partial z^2}(\Omega \Psi) \quad ,$$

where

$$\Omega = (\mu + \alpha)(\beta + \gamma - \varepsilon)\nabla^2 - 4\alpha \quad , \quad \Phi = (\mu + \alpha)\nabla^2[(\beta + 2\gamma)\nabla^2 - 4\alpha] \, ,$$

and we obtain, from the system of the elastostatic equations, the following equation in terms of the second generalized function of Love

$$16\alpha^2\mu(h^2\nabla^2-1)(l^2\nabla^2-1)\Psi(r,z)+Y_z(r,z)=0 \qquad (3.36)$$

J. Stefaniak [57] derived the analogous functions from the vector functions of N. Sandru.

Another method of solving the equations of the axially symmetric problems consists in the introduction of elastic potentials [56]. We shall explain this method shortly on the example of the first axially symmetric problem. Let us express the displacements $\underline{u}=(u_r, 0, u_z)$ and the rotations $\underline{\varphi}=(0, \varphi_\theta, 0)$ by means of the potentials Φ, Ψ, ϑ

$$u_r=\frac{\partial\Phi}{\partial r}+\frac{\partial^2\Psi}{\partial r\partial z}, \quad u_z=\frac{\partial\Phi}{\partial z}-\frac{\partial^2\Psi}{\partial r^2}-\frac{1}{r}\frac{\partial\Psi}{\partial r}, \qquad (3.37)$$
$$\varphi_\theta=\frac{\partial\vartheta}{\partial r}.$$

Substituting eqs. (3.37) intc the equations of elastostatics we obtain the system of two equations

$$\nabla^2\nabla^2\Phi=0, \qquad \nabla^2(l^2\nabla^2-1)=0. \qquad (3.38)$$

The functions Φ and ϑ are connected by the relations

$$\nabla^2\Phi-\frac{2\mu}{\lambda+2\mu}\frac{\partial}{\partial z}(l^2\nabla^2-1)\vartheta=0,$$
$$\frac{\partial}{\partial z}\nabla^2\Phi+\frac{2\mu}{\lambda+\mu}\left(\nabla^2-\frac{\partial^2}{\partial z^2}\right)(l^2\nabla^2-1)\vartheta=0. \qquad (3.39)$$

The function Ψ will be determined by the equation

$$\nabla^2\Psi=-\frac{1}{2\alpha^2}[(\gamma+\varepsilon)\nabla^2-4\alpha]\vartheta. \qquad (3.40)$$

Exactly the same procedure can be applied in the second axially-symmetric problem. This purpose can be attained also by different methods, as the application of Neuber's function, or, finally by the direct integration of the system of differential equations by means of the Hankel integral transform or Hankel-Fourier transform. A number of special problems, concerning the state of stress in an elastic semi-space has been solved (the generalized Boussinesq problem).

We conclude the review of the micropolar elastostatics devoting a few words to the general variational theorems, the theorem on reciprocity, and so on. The theorems of this type turned out to be easy to extend on the micropolar theory of elasticity by the addition of the corresponding terms connected with the work of moments and the moment stresses.

The principle of the virtual work of the virtual displacements δu_i and the virtual rotations $\delta\varphi_i$ takes the form

$$(3.41) \quad \int_V (X_i \delta u_i + Y_i \delta\varphi_i)\, dV + \int_A (p_i \delta u_i + m_i \delta\varphi_i)\, dA = \delta W_\varepsilon ,$$

where

$$W_\varepsilon = \int (\mu \gamma_{(ij)} \gamma_{(ij)} + \alpha \gamma_{<ij>} \gamma_{<ij>} + \frac{\lambda}{2} \gamma_{kk} \gamma_{nn} + \ldots)\, dV .$$

The minimum of potential energy theorem can be derived from the principle of virtual work

$$\delta\Gamma = 0 \quad , \tag{3.42}$$

where

$$\Gamma = W_\varepsilon - \int_V (X_i u_i + Y_i \varphi_i) dV - \int_{A_\sigma} (p_i u_i + m_i \varphi_i) dA \; .$$

Here A_σ denotes the part of the surface bounding the body where the loadings are prescribed.

In the classical elastostatics an important role is played by the theorem of minimum complementary energy. Here in the micropolar elastostatics we have the following variational principle

$$\delta\Pi = 0 \quad , \tag{3.43}$$

where

$$\Pi = W_\sigma - \int_{A_u} (p_i u_i + m_i \varphi_i) dA \quad ,$$

and

$$W_\sigma = \int_V \left(\mu' \sigma_{(ij)} \sigma_{(ij)} + \varepsilon' \sigma_{<ij>} \sigma_{<ij>} + \frac{\lambda'}{2} \sigma_{kk} \sigma_{nn} + \; . \; . \; . \right) dV \; .$$

A_u denotes the part of the bounding surface for which the displacements and rotations are prescribed.

N. Sandru [20] derived the reciprocity theorem

$$\begin{aligned} \int_V (X_i u_i' + Y_i \varphi_i') dV + \int_A (p_i u_i' + m_i \varphi_i') dA &= \\ = \int_V (X_i' u_i + Y_i' \varphi_i) dV + \int_A (p_i' u_i + m_i' \varphi_i) dA \; , \end{aligned} \tag{3.44}$$

and the particular case of the theorem, namely the generalized formulae of Somigliano. Much attention has been devoted to the uniqueness of solution and the existence theorem (M. Hlavacek [58], D. Iesan [59]).

The research work on the theory of dislocations in the Cosserat continuum needs a separate treatment. W. Günther was the first [6] who noticed the importance of the micropolar elasticity for the theory of dislocations. Certain concepts concerning the dislocations were discussed in the paper of W. D. Claus and A. C. Eringen [60] and in the paper of S. Minagawa [61] .

The theory of anisotropic Cosserat continuum was investigated in papers by S. Kessel [62] and D. Iesan [63] .

The theory of the Cosserat continuum is well developed. At the present day it constitutes a complete superstructure of the classical theory of elasticity. However, the complete experimental verification of the theory is still lacking. The material constants μ , λ , α , β , γ , ε have not been determined for particular materials. We only know the order of these constants and the mutual ratios of the six constants. Thus we have here the extreme case when the theory outdistances the experiments.

In the further development of the Cosserat mechanics the main role should be played by the experimental research. The role of the theoreticians is here already exhausted.

REFERENCES

[1] Voigt W.: "Theoretische Studien über die Elastizitätverhältnisse der Kristalle", Abh. Ges. Wiss. Göttingen 34, (1887).

[2] Cosserat E. and Cosserat F.: "Théorie des corps déformables", A. Herrman, Paris, (1909).

[3] Truesdell C. and Toupin R. A.: "The classical field theories", Encyclopaedia of Physics 3, N°1, Springer Verlag, Berlin, 1960.

[4] Grioli G.: "Elasticité asymmetrique", Ann. di Mat. Pura et Appl. Ser. IV, 50 (1960).

[5] Mindlin R. D. and Tiersten H. F.: "Effects of couple stresses in linear elasticity", Arch. Mech. Analysis, 11 (1962), 385.

[6] Günther W.: "Zur Statik und Kinematik des Cosseratschen Kontinuums", Abh. Braunschweig, Wiss. Ges. 10 (1958), 85.

[7] Schaefer H.: "Versuch einer Elastizitätstheorie des zweidimensionalen Cosserat-Kontinuums, Misz. Angew. Mathematik Festschrift Tollmien, Berlin, (1962), Akademie Verlag.

[8] Neuber H.: "On the general solution of linear-elastic problems in isotropic and anisotropic Cosserat-continua", Int. Congress IUTAM, München, 1964.

[9] Kuvshinskii E. V. and Aero A. L.: "Continuum theory of asymmetric elasticity" (in Russian), Fizika Tvordogo Tela, 5 (1963).

[10] Palmov N.A.: "Fundamental equations of the theory of asymmetric elasticity" (in Russian), Prikl. Mat. Mekh. 28 (1964), 401.

[11] Eringen A.C. and Suhubi E.S.: "Nonlinear theory of simple microelastic solids", Int. J. of Engng. Sci. I, 2, 2 (1964), 189; II, 2, 4 (1964), 389.

[12] Stojanovic R.: "Mechanics of Polar Continua", CISM, Udine (1970).

[13] Nowacki W.: "Theory of micropolar elasticity", J. Springer, Wien (1970).

[14] Nowacki W.: "Propagation of rotation waves in asymmetric elasticity", Bull. Acad. Polon. Sci. Sér. Sci. Techn. 16, 10 (1968), 493.

[15] Ignaczak J.: "Radiation conditions of Sommerfeld type for elastic materials with microstructure", Bull. Acad. Polon. Sci. Sér. Sci. Techn. 17, 6 (1970), 251.

[16] Nowacki W. and Nowacki W.K.: "The generation of waves in an infinite micropolar elastic solid", Proc. Vibr. Probl. 10, 2 (1969), 170.

[17] Nowacki W.: "On the completeness of potentials in micropolar elasticity", Arch. Mech. Stos., 21, 2, (1969), 107.

[18] Galerkin B.: "Contributions à la solution générale du probleme de la théorie de l'élasticité dans le cas de trois dimensions", C.R. Acad. Sci. Paris, 190 (1970), 1047.

[19] Iacovache M.: "O extindere a metodei lui Galerkin pentru sistemul ecuatiilor elasticitatii", Bull. Stiint. Acad. Rep. Pop. Romane. Ser. A 1 (1949), 593.

[20] Sandru N.: "On some problems of the linear theory of asymmetric elasticity", Int. J. Eng. Sci. 4, 1, (1966), 81.

[21] Stefaniak J.: "Generalization of Galerkin's functions for asymmetric thermoelasticity", Bull. de l'Acad. Polon. Sci. Sér. Sci. Techn. 16, 8, (1968), 391.

[22] Olesiak Z.: "Stress differential equations of the micropolar elasticity", Bull. Ac. Polon. Sci. Sér. Sci. Techn. 18, 5, (1970), 172.

[23] Eringen A.C.: "Theory of micropolar elasticity", Fracture, 2 (1968), Academic Press, New York.

[24] Smith A.C.: "Waves in micropolar elastic solids", Int. J. Engng. Sci. 10, 5 (1967), 741.

[25] Parfitt V.R. and Eringen A.C.: "Reflection of plane waves from the flat boundary of a micropolar elastic halfspace", "Report N° 8-3, General Technology Corporation, (1966).

[26] Stefaniak J.: "Reflection of a plane longitudinal wave from a free plane in a Cosserat medium", Arch. Mech. Stos. 21, 6 (1969), 745.

[27] Kaliski S., Kapelewski J. and Rymarz C.: "Surface waves on an optical branch in a continuum with rotational degrees of freedom", Proc. Vibr. Probl. 9, 2, (1968), 108.

[28] Nowacki W. and Nowacki W.K.: "Propagation of monochromatic waves in an infinite micropolar elastic plate", Bull. Acad. Polon. Sci., Sér. Sci. Techn. 17, 1 (1969), 29.

[29] Nowacki W. and Nowacki W.K.: "The plane Lamb problem in a semi-infinite micropolar elastic body" Arch. Mech. Stos. 21, 3, (1969), 241.

[30] Eason G.: "Wave propagation in a material with microstructure", Proc. Vibr. Probl. 12, 4, (1971), 363.

[31] Eason G.: "The displacement produced in a semi-infinite linear Cosserat continuum by an impulsive force", Proc. Vibr. Probl., 11, 2, (1970), 199.

[32] Achenbach J.D.: "Free vibrations of a layer of micropolar continuum" Int. J. Engng. Sci. 10, 7 (1969), 1025.

[33] Nowacki W. and Nowacki W.K.: "Propagation of elastic waves in a micropolar cylinder", Bull. Acad. Polon. Sci., Sér. Sci. Techn. I-17, 1 (1969), 49.

[34] Nowacki W. and Nowacki W.K.: "The axially symmetric Lambs problem in a semi-infinite micropolar elastic solid", Proc. Vibr. Probl. 10, 2 (1969), 97.

[35] Iesan D.: "On the linear theory of micropolar elasticity", Int. J. Eng. Sci. 7 (1969), 1213.

[36] Graffi D.: "Sui teoremi di reciprocità nei fenomeni non stazionari", Atti Accad. Sci. Bologna, 10, 2, (1963).

[37] Schaefer H.: "Das Cosserat-Kontinuum", ZAMM, 47, 8 (1967), 485.

[38] Neuber H.: "Über Probleme der Spannungskonzentration im Cosserat-Körper", Acta Mechanika, 2, 1 (1966), 48.

[39] Neuber H.: "Die schubbeanspruchte Kerbe im Cosserat-Körper", ZAMM, 47, 5, (1967), 313.

[40] Neuber H.: "On the effects of stress concentration in Cosserat continua", IUTAM-Symposium (1967),

Freudenstadt-Stuttgart, Mechanics of generalized continua (1968), 109.

[41] Cowin S.C.: "Stress functions for elasticity", Int. J. Solids Structures, 6, (1970), 389.

[42] Schaefer H.: "Die Spannungsfunktionen eines Kontinuum mit Momentenspannungen", Bull. de l'Acad. Sci. Pol., Sér. Sci. Techn. I, 15, 1 (1967), 63; II, 15, 1, (1967), 485.

[43] Mindlin R.D.: "Influence of couple-stresses on stress concentrations", Exper. Mech. 3, 1, (1963), 1.

[44] Savin G.N.: "Foundation of couple stresses in elasticity" (in Ukrainian), Kiev, (1965).

[45] Savin G.N.: "Stress concentration in the neighbourhood of holes" (in Russian), Kiev (1966).

[46] Savin G.N. and Nemish J.N.: "Stress concentration in couple-stress elasticity" (in Russian), Prikl. Mech. 4, 12 (1968), 1.

[47] Carlson D.E.: "Stress functions for plane problems with couple-stresses", ZAMP, 17, 6, (1966), 789.

[48] Mindlin R.D.: "Representation of displacements and stresses in plane strain with couple-stresses", IUTAM Symposium (1963), Tbilisi (1954), 256.

[49] Muki R. and Sternberg E.: "The influence of couple-stress concentrations in elastic solids", ZAMP,, 16, 5, (1965), 611.

[50] Kaloni P.N. and Ariman T.: "Stress concentration effects in micropolar elasticity", ZAMP, 18, 1, (1967), 136.

[51] Kessel S.: "Die Spannungsfunktionen des Cosserat-Kontinuum", ZAMM, 47, 5 (1967), 329.

[52] Nowacki W.: "Plane problems of micropolar elasticity", Bull. Acad. Polon. Sci., Sér. Sci. Techn. 19, 6, (1971), 237.

[53] Nowacki W.: "The "second" plane problem of micropolar elasticity", Bull. Acad. Polon. Sci., Sér. Sci. Techn. 18, 11 (1970), 525.

[54] Suchar M.: "Stress functions in the "second" plane problem of micropolar elasticity", Bull. Acad. Polon. Sci. 20, (1972) (in print).

[55] Nowacki W.: "Generalized Love's functions in micropolar elasticity", Bull. Acad. Polon. Sci. Sér. Sci. Techn., 17, 4, (1969), 247.

[56] Nowacki W.: "Axially symmetric problem in micropolar elasticity", Bull. Acad. Polon. Sci., Sér. Sci. Techn. 19, 7/8, (1971), 317.

[57] Stefaniak J.: "Solving functions for axi-symmetric problems in the Cosserat medium", Bull. de l'Acad. Polon. Sci., Sér. Sci. Techn. 18, 9, (1970), 387.

[58] Hlavacek M.: "The first boundary-value problem of elasticity theory with couple stresses I. Cosserat Continuum", Bull. Acad. Polon. Sci., Sér. Sci. Techn. 18, 2, (1970), 75.

[59] Iesan D.: "Existence theorems in micropolar elastostatics", Int. J. Eng. Sc. 9, (1971), 59.

[60] Claus W. D. and Eringen A. C.: "Three dislocation concepts and micromorphic mechanics. Developments in Mechanics" Vol. 6. Proceedings of the 12th Midwestern Mechanics Conference.

[61] Minagawa S.: "On the force exerted upon continuously distributed dislocations in a Cosserat continuum", Phys. Stat. Sol., 39 (1970), 217.

[62] Kessel S.: "Lineare Elastizitätstheorie der anisotropen Cosserat-Kontinuums", Abh. der Braunschw. Wiss. Gesellschaft 16, (1964), 1.

[63] Iesan D.: "Thermal stresses in the generalized plane strain of anisotropic elastic solids", Bull. de l'Ac. Pol. Sc., Sér. Sci. Techn. 18, 6, (1970), 227.

G. GRIOLI

LINEAR MICROPOLAR MEDIA WITH CONSTRAINED ROTATIONS

In the mechanics of non-polar continuous media all torques are assumed to be the moments of forces; it means that there are neither body couples nor couple stresses.

In more general formulations of the continuous media theory embodying a more detailed molecular picture of matter, there may be also couple-stresses or hyper-stresses which correspond to the more general kinds of deformations. Such deformations are said to be hyperdeformations.

Since any system of forces is equivalent to the resultant force acting at the given point and to the resultant torque, it seems to be that the more general concept of the continuous medium is that of Cosserat continuum of micropolar medium.

Assuming that the couple-stresses are present, we have to take into account also local measures of rotations on which the couple stresses do the work. These rotations can be either uniquely determined by the displacement field or independent of the displacements. In the former case we deal with the micropolar continuum with constrained rotations and in the latter case the micropolar continuum with free rotations is taken into account.

The concept of local rotation is defined for nonpolar continuum; it is the rotation we are going to deal with in what follows, investigating the micropolar continuum with constrained rotations. The micropolar continuum with constrained rotations constitutes an interesting special case of the Cosserat continuum in which the displacement field determines all geometric and kinematic properties of the deformation.

The formulation of the theory of continuous media in which the stresses as well as the thermodynamical potentials depend also on the second-order and higher-order derivatives of the displacement field is possible. Using such approach we shall arrive at a more general form of the stress relations than in the classical case. For example, in the theories with couple stresses we have to take into account the second-order derivatives of the displacement field. At the same time the suitable compatibility conditions have to be satisfied for each correctly stated theory.

Little attention was drawn to the Cosserats' work until recent years, when the behaviour of the non-polar media has received a great deal of attention, including the theories of couple stresses with constrained rotations (1) [1], [2], [3] , [4], ... [48] .

In the lecture we are to investigate the connection between linear and non-linear theories of non-polar continua.

We shall start with the short survey of the basic equations and theorems for the linear micropolar continuum; that will enable us to give some new remarks on this topic.

We shall confine ourselves to the pure mechanical theory of micropolar continuous media: the thermomechanical formulation of the micropolar theory can be analogously obtained as in the non-polar case.

(1) The problem of free rotations has been considered much more than that with constrained rotations. The bibliography concerns both kinds of problems.

1. Geometry and Kinematics

Let be given the material continuum in two distinct configurations, C and C'. Let P be a point in C and denote by P' the point in C', which is occupied by the same particle. The coordinates of points P and P' with respect to the fixed Carthesian reference frame τ, will be denoted by y_i and x_i, respectively.

For the time being, let to each point P be attached a rigid material particle c with the centre of mass at P. Let us denote by c' the position of c in the configuration C', and by P' the centre of mass of the particle in this configuration.

Thus we can treat P as a rigid body whose position in C' is denoted by P'. Let Q be a material point in c which is carried to the place Q' in c'. Moreover, let us denote by η_i the coordinates of the point Q with respect to the rectangular Carthesian coordinate frame having the origin at the point P and coordinate axes parallel to those of the coordinate system τ.

In the linear case

$$P'Q' = \underline{\omega} \times PQ + PQ , \tag{1}$$

where $\underline{\omega}$ denotes the rotation of c due to the transformation from C to C'.

Let us denote by $e_{rs\ell}$ the Ricci's tensor, by u_r the components of the displacement and by ε_{rs} the linearized strain measure. The following well known relations hold

$$\varepsilon_{rs} = \frac{1}{2}\left(u_{r,s} + u_{s,r}\right), \qquad \omega_r = \frac{1}{2}\, e_{rpq}\, u_{q,p}\,, \tag{2}$$

where the partial derivatives with respect to y_i are indicated by a comma in front of the corresponding index.

Denoting by δ_{rs} the Kronecker delta, from (1) and (2) we shall obtain

$$\begin{aligned}(dQ')_r = &\left[\varepsilon_{rs} + e_{rps}\,\omega_p + e_{rpq}\,\omega_{p,s}\,\eta_q + \delta_{rs}\right] dy_s + \\ &+ \left[e_{rpq}\,\omega_p + \delta_{rq}\right] d\eta_q\,.\end{aligned} \tag{3}$$

where dQ is a material element in C and dQ' is its image in C'.

It follows that the transformation from C to C' (the mapping of C onto C') is characterized by the metric

$$\begin{aligned}ds'^2 = &\left[\delta_{rs} + 2\left(\varepsilon_{rs} + e_{rpq}\,\omega_{p,s}\,\eta_q\right)\right] dy_r\, dy_s + \\ &+ \delta_{rs}\, d\eta_r\, d\eta_s + 2\left[\delta_{rs} + \varepsilon_{rs} + e_{spq}\,\omega_{p,r}\,\eta_q\right] dy_r\, d\eta_s\,.\end{aligned} \tag{4}$$

It is easy to establish that the extension of a linear material element can be obtained from the formula

$$ds'^2 - ds^2 = 2\left[\left(\varepsilon_{rs} + e_{rpq}\,\omega_{p,s}\,\eta_q\right)\right]\left[dy_r\, dy_s + dy_s\, d\eta_r\right] \tag{5}$$

We can prove that the conditions $\varepsilon_{rs} = 0$, $\omega_{rs} = 0$ are both necessary and sufficient for the displacement to be rigid (however, we may observe that if $\varepsilon_{rs} = 0$ then $\omega_{rs} = 0$). Thus the matrices ε_{rs} , ω_{rs} determine the deformation of the continuum. The work done by the internal forces in the deformation from C' to $C' + \delta C'$ will be given by

$$\delta \ell^{(i)} = Y_{rs} \delta \varepsilon_{rs} + \varphi_{rs} \delta \mu_{rs} , \tag{6}$$

and the rate at which the stresses do work is given by the stress power

$$\mathcal{P}^{(i)} = Y_{rs} \dot{\varepsilon}_{rs} + \varphi_{rs} \dot{\mu}_{rs} , \tag{7}$$

where we have denoted

$$\mu_{rs} = \omega_{r,s} = \frac{1}{2} e_{rpq} u_{q,ps} = e_{rpq} \varepsilon_{qs,p} . \tag{8}$$

Note that the components μ_{rs} satisfy the following equality

$$I(\mu) = \mu_{rr} = 0 . \tag{9}$$

Since the quadratic form (4) has to represent the Euclidean metric, we shall obtain the following necessary and sufficient conditions of compatibility for the tensors ε_{rs} and μ_{rs} :

$$e_{rpl}\, e_{sqm}\, \varepsilon_{pq,lm} = 0\ , \qquad \mu_{rs} - e_{rpq}\, \varepsilon_{qs,p} = 0\ . \tag{10}$$

The quantities with components Υ_{rs} , φ_{rs} represent the stress tensor and the couple stress tensor, respectively. By virtue of (6) and (7) we can assume that they depend on ε_{rs} and μ_{rs} .

A micropolar continuum is said to be a simple linear micropolar continuum if the stresses, couple stresses, free energy etc. depend on ε_{rs} , μ_{rs} . The constitutive equations in such continuum depend on the history of the local configuration or they may be elastic, quasi-elastic, hyperelastic.

The constitutive equations are frame indifferent. For hyperelastic materials there exists the strain energy function, whose arguments are ε_{rs}, μ_{rs} and the temperature; the quantities Υ_{rs} and φ_{rs} can be derived from this function.

To define the dynamical process we have to attribute to each material point a density γ and a symmetric structural matrix A_{rs}, which are functions of the particle P, but independent of the time coordinate. The following expressions for the kinetic energy T and the intrinsic moment of momentum can then be written down

$$T = \frac{1}{2} \int_C \gamma \left[\dot{u}^2 - I_1(A)\,\dot{\omega}^2 - A_{pq}\,\dot{\omega}_p\dot{\omega}_q\right] dC\ , \tag{11}$$

$$K_r^{(p)} = \int_C \gamma\, e_{rln}\, e_{npq}\, A_{lq}\, \omega_p\, dC\ . \tag{12}$$

2. Basic Equations

Equality (9) implies that the stresses remain in general undetermined. From this follows that by virtue of (9), the relations (6) and (7) can be put in the form

(6')
$$\delta\ell^{(i)} = Y_{(rs)}\,\delta\varepsilon_{rs} + N_{rs}\,\delta\mu_{rs} \;,$$

(7')
$$\mathcal{P}^{(i)} = Y_{(rs)}\,\dot{\varepsilon}_{rs} + N_{rs}\,\dot{\mu}_{rs} \;,$$

where

(13)
$$N_{rs} = \varphi_{rs} - \delta_{rs}\,\varphi_{33} \;.$$

It is clear that the expressions for $\delta\ell^{(i)}$ and $\mathcal{P}^{(i)}$ do not depend on all components of the stress tensor and couple stress tensor but only on the symmetric part of the matrix Y_{rs} and on the eight components N_{rs}, $(N_{33}=0)$. Hence, through the use of the well known thermodynamical approach, for various linear micropolar materials (e.g. elastic, hyperelastic) we obtain the definite positive quadratic form of the arguments $\varepsilon_{rs}, \mu_{rs}, \dot{\varepsilon}_{rs}, \dot{\mu}_{rs}$, not containing the terms $r = s = 3$ and such that [2]

(2) $W^{(1)}, W^{(2)}$ may depend on other parameters, as, for example co-ordinates, etc.

$$\left.\begin{aligned} Y_{r s} &= -\frac{\partial W^{(1)}}{\partial \varepsilon_{r s}} - \frac{\partial W^{(2)}}{\partial \dot{\varepsilon}_{r s}}, \\ N_{r s} &= -\frac{\partial W^{(1)}}{\partial \mu_{r s}} - \frac{\partial W^{(2)}}{\partial \dot{\mu}_{r s}}, \quad E = -\frac{\partial W^{(1)}}{\partial \vartheta} \end{aligned}\right\} \tag{14}$$

where E is the specific entropy and ϑ is the absolute temperature.

The constitutive equations (14) will hold if the reference configuration is an unstressed state of the material. For hyperelastic materials $W^{(2)}$ is equal to zero.

The dynamic equations are

$$Y_{r s, s} = \gamma\left(F_r - \ddot{u}_r\right), \qquad (\text{in } C), \tag{15}$$

$$N_{r s, s} + \delta_{r s} \varphi_{33, s} + e_{rpq} Y_{qp} = \gamma\left[M_r + A_{rp} \ddot{\omega}_p - I_1(A) \ddot{\omega}_r\right], \quad (\text{in } C), \tag{16}$$

$$Y_{r s} n_s = f_r, \qquad \left(N_{r s} + \delta_{r s} \varphi_{33}\right) n_s = m_r, \quad (\text{on } \sigma), \tag{17}$$

where all symbols have the well known meaning and where n_s are components of the interior unit normal vector on the boundary σ of C.

The antisymmetric part of Y_{rs} is determined by Eqs. (16)

$$\begin{aligned} Y_{[r s]} = \frac{1}{2} e_{r s p}\Big\{ & \left[N_{p \ell} + \delta_{p \ell} \varphi_{33}\right]_{, \ell} - \\ & -\gamma\left[M_p + A_{p \ell} \ddot{\omega}_{\ell} - I_1(A) \ddot{\omega}_p\right]\Big\}. \end{aligned} \tag{18}$$

Thus the system of equations (15), (16), (17) may be expressed as follows

$$Y_{(rs),s} + \frac{1}{2} e_{rsp}\left[N_{pl,l} - \gamma\left(M_p + A_{pl}\ddot{\omega}_l - I_1(A)\ddot{\omega}_p\right)\right]_{,s} =$$

$$= \gamma\left(F_r - \ddot{u}_r\right), \quad (\text{in } C) \tag{19}$$

and the corresponding boundary conditions are

$$\left\{Y_{(rs)} + \frac{1}{2} e_{rsp}\left[\left(N_{pl} + \delta_{pl}\varphi_{33}\right)_{,l} - \gamma\left(N_p + A_{pl}\ddot{\omega}_l - \right.\right.\right.$$

$$\left.\left.\left. - I_1(A)\ddot{\omega}_p\right)\right]\right\} n_s = f_r, \quad (\text{on } \sigma) \tag{20}$$

The component φ_{33} does not enter the equation (19), however it is present in the boundary conditions. Eqs. (19) are differential equations of the fifth order in general case; for hyperelastic materials $W^{(2)}$ is equal to zero and (19) become equations of the fourth order.

If C is in an equilibrium state, from which the external loads are absent (free equilibrium state) and if the body is homogeneous, then C will be an unstressed configuration (natural state of equilibrium). In this case, from (15), (17.1), we conclude that

$$\int_C Y_{rs}\, dC = -\left[\int_C F_r y_s\, dC + \int_C f_r y_s\, d\sigma\right]. \tag{21}$$

For the homogeneous body Y_{rs} are constant. Because of $F_r \equiv f_r \equiv 0$ from (21) follows that $Y_{rs} = 0$. Analogously, since $M_r \equiv m_r \equiv 0, \ddot{\omega}_r \equiv 0$, by virtue of (16), (17.2) we arrive at $N_{rs} + \delta_{rs}\varphi_{33} \equiv \varphi_{rs} \equiv 0$.

For the linear micropolar hyperelastic continuum we have to put $W^{(2)} \equiv 0$ into (14), reducing the form of the constitutive equations to the following one

$$Y_{(rs)} = -\frac{\partial W^{(1)}}{\partial \varepsilon_{rs}}, \quad N_{rs} = -\frac{\partial W^{(1)}}{\partial \mu_{rs}}, \tag{22}$$

where $W^{(1)}$ does not depend on μ_{33}. For isotropic materials the function $W^{(1)}$ has the form (13)

$$W^{(1)} = W'(\varepsilon) + W''(\mu), \tag{23}$$

where $W'(\varepsilon)$ has the same form as in the classical theory of elasticity

$$W'(\varepsilon) = \frac{1}{2}\left[(\tau + 2\nu)\, I_1^2(\varepsilon) - 4\nu\, I_2(\varepsilon)\right] \tag{24}$$

and where

$$W''(\mu) = \frac{1}{2}\left[\sum_{r,s}{}^{*}(B\mu_{rs} + C\mu_{sr})\mu_{rs} + (B - C)(\mu_{11} + \mu_{22})^2\right]. \tag{25}$$

The coefficients ν, τ, B, C have to satisfy the inequalities

$$\nu > 0, \quad 3\tau + 2\nu > 0, \quad B > 0, \quad |C| < B \tag{26}$$

The sum Σ^* in (25) does not contain the term which corresponds to the indices $r = s = 3$. At the same time we have

$$A_{rs} = \delta_{rs}\, \varrho^2 \tag{27}$$

where ϱ is a known structural parameter.

3. Uniqueness Theorem

A uniqueness theorem for the solution of the set of equations (14), (15), (16), (17) with suitable boundary conditions follows from the fact that for hyperelastic materials $\left[W^{(2)} = 0\right]$ we have

$$\int_C W^{(1)} dC + T = L^{(e)} + \int_C W_0^{(1)} dC + T_0 , \tag{28}$$

where $L^{(e)}$ is the work done by the external loads from the initial to the actual time instant t.

Let us suppose that there are given two solutions $u_r^{(i)}, \omega_r^{(i)}, (i=1,2)$ of the problem which correspond to the same initial conditions and to the same body forces and body couples. Putting

$$u_r = u_r^{(1)} - u_r^{(2)}, \quad \omega_r = \omega_r^{(1)} - \omega_r^{(2)}, \quad f_r = f_r^{(1)} - f_r^{(2)}, \quad m_r = m_r^{(1)} - m_r^{(2)}, \tag{29}$$

we arrive at $u_r(y_i, 0) = \omega_r(y_i, 0) = 0$ and, moreover, for u_r, ω_r we have $W_0^{(1)} = T_0 = 0$. From (28) follows that

$$\int_C W^{(1)}(u,\omega)\, dC + T(\dot{u},\dot{\omega}) = \int_0 dt \int_\sigma (f_r u_r + m_r \omega_r)\, d\sigma . \tag{30}$$

Because the forms $W^{(1)}$ and T are positive definite it is sufficient condition for uniqueness that the conditions

$$f_r \dot{u}_r + m_r \dot{\omega}_r = 0 , \tag{31}$$

are satisfied on the boundary for any $t \geq 0$. For example, on the boundary σ either the vectors u_r, ω_r or the external surface tractions and couples can be known.

4. Wave Propagation

The problems of uniqueness, stress functions, wave propagation, potential of dilatation and rotation may be found in [15]. Here we shall confine ourselves to some remarks on the wave propagation problem for the linear isotropic hyperelastic material. In this case the equations (22)... (27) hold, and Eq. (19), if the body loads are equal to zero, gives

$$
(32) \qquad \begin{aligned} &(\tau+\nu)u_{s,sr} + \nu u_{r,ss} + \frac{B}{4}\left(u_{s,sr} - u_{r,ss}\right)_{,ii} = \\ &= \gamma\left[\ddot{u}_r - \frac{\varrho^2}{4}\left(\ddot{u}_{r,ss} - \ddot{u}_{s,sr}\right)\right]. \end{aligned}
$$

It is clear that if $(u_{rs} - u_{sr} = 0)$ equations (32) will have the classical form, from which we shall obtain the classical expression for the velocity of propagation $\sqrt{\frac{\tau+2\nu}{\gamma}}$.

For solenoidal displacement field $(u_{r,r} = 0)$ equations (32) reduce to the form

$$
(33) \qquad \left[\nu u_r - \frac{B}{4}u_{r,ss}\right]_{,ii} = \gamma\left[\ddot{u}_r - \frac{\varrho^2}{4}\ddot{u}_{r,ss}\right].
$$

To analyze the problem of the speed propagation of the surface on which the fourth derivatives suffer a jump discontinuity, we have to evaluate a jump of the moment of momentum in the dynamic equations. A well known procedure leads to the following expression for the speed propagation

$$V = \sqrt{\frac{B}{\gamma \varrho^2}} \tag{34}$$

which is independent of ν .

If the value of V has to be close to the corresponding value for the non-polar case, then the value ϱ^2 has to be close to $\frac{B}{\nu}$. Assuming that the particles are of the cubical or spherical shape, the idea of the size of the particles follows from the measurement of B and ν .

5. On the Statics of Linear Micropolar Continua

In what follows we shall assume that the problem is static and the body may not be isotropic. The basic equations are

$$\left.\begin{aligned} &(35) \qquad Y_{rs,s} = \gamma F_r \\ &(36) \qquad N_{rs,s} + \delta_{rs}\varphi_{33,s} + e_{rpq}Y_{qp} = \gamma M_r \end{aligned}\right\} \text{(in C)},$$

$$(37) \qquad Y_{rs}\, n_s = f_r\,, \quad N_{rs} n_s = m_r\,, \qquad (\text{on } \sigma).$$

Besides the above given equations, we have to take into account the constitutive equations which, for the hyperelastic materials are given by (22).

Using the classical procedure we can prove some theorems, e. g. the theorems of Betti or Clapeyron also in the non-polar case. The classical Menabrea's theorem also holds in the non-polar case, but the form of this theorem slightly differs from the well known formulation because of the indeterminacy of the stresses [29]. Now we are to give a more general interpretation of the classical Menabrea's theorem which enables us to establish an existence theorem and to construct an integration method.

Since the quadratic form $W^{(1)}$ is positive definite we can deduce from (22) the values of ε_{rs}, μ_{rs} as the functions of Y_{rs}, N_{rs}. Thus the definite positive quadratic form $W^*(Y,N)$ exists, being the function of the fifteen variables Y_{rs}, N_{rs} and for which the following equalities hold.

$$\left.\begin{aligned} \varepsilon_{rs} &= -\frac{\partial W^*}{\partial Y_{(rs)}} \\ \mu_{rs} &= -\frac{\partial W^*}{\partial N_{rs}} \end{aligned}\right\} \tag{38}$$

Let us assume that the surface tractions are known on a part of the boundary, and on the other part σ'' the values of the displacements are prescribed $u_r = \bar{u}_{rs}$, $\omega_r = \bar{\omega}_r$.

Putting

$$A = \int_C W^*(Y,N)\,dC - \int_{\sigma''} (\Phi_r \bar{u}_r + M_r \bar{\omega}_r)\,d\sigma'' \tag{39}$$

where Φ_r, M_r are unknown forces and couples due to the constraints, we can formulate the following theorem:

The real values of stresses and couple stresses will ensure the minimum value of A in the class of all stresses and couple stresses which are in equilibrium with the body loads in C and the known surface tractions on σ'' if Φ_r and M_r belong to the class of all possible reaction forces and couples which are in equilibrium with the known external forces and couples.

Conversely, if A admits a minimum Y'_{rs} N'_{rs} for the prescribed $\bar{u}_r$, $\bar{\omega}_r$ on σ'' in the class of all solutions of the equations (35), (36) in C and (37) on σ', then the strain and the rotation $\varepsilon_{rs}\,\omega_{rs}$ determined by Eq. (38) will satisfy the compatibility conditions (10), and will ensure the existence of the displacement field u_r and the rotation field ω_r satisfying the conditions $u_r = \bar{u}_{r}$, $\omega_r = \bar{\omega}_r$ almost everywhere on σ''. In the author's opinion, an analogous theorem may also hold when on σ'' the unilateral constraints are given.

Before the conclusion of the lecture we are to make some remarks on the connection between the linear and non-linear theories in the static case. We are to prove that the rigid rotation which is supposed to be indetermined in the linear theory, is actually determined if the linear theory is treated as the first approximation of the general theory.

Denoting by X_{rs} and Ψ_{rs} the stresses and couple stresses respectively in the deformed configuration, and putting

$$\text{(40)}\quad \begin{cases} X_{rs} = \dfrac{1}{\mathfrak{D}}\, x_{r,l}\, x_{s,m}\, Y_{lm}\,, \qquad \Psi_{rs} = \dfrac{1}{\mathfrak{D}}\, x_{r,l}\, x_{s,m}\, \varphi_{lm}\,, \\[2ex] \mathfrak{D} = \left\| x_{r,s} \right\| = \mathrm{Det.} \left| x_{r,s} \right|, \end{cases}$$

we shall obtain the following form of the equilibrium equations for a micropolar continuum subjected to the finite deformations

$$\text{(41)}\quad \begin{cases} \left(x_{r,l}\, Y_{ls}\right)_{,s} = \lambda \gamma F_r\,, \\[1ex] \left(x_{r,l}\, \varphi_{ls}\right)_{,s} + e_{rpt}\, x_{t,l}\, x_{p,m}\, Y_{lm} = \lambda \gamma M_r\,, \end{cases} \qquad (\text{in } C),$$

$$\varkappa_{r,l}\, Y_{ls}\, n_s = \lambda f_r\,, \quad \varkappa_{r,l}\, \varphi_{ls}\, n_s = \lambda m_r\,, \tag{42}$$

where λ denotes a multiplicative parameter of external loads.

Let the functions u_r, Y_{rs} etc. have derivatives with respect to λ up to the second order at $\lambda = 0$. If the functions u_{rs}, Y_{rs}, have derivatives of an arbitrary order with respect to λ at $\lambda = 0$,then it will be possible to express the solution of the equilibrium problem for finite deformations by means of the power series with respect to the parameter λ . In this case the solution of the non-linear problem may be reduced to the solution of a system of linear problems, the number of such solutions being finite in practice.

For any function g of an argument λ we shall put

$$g^{(i)} = \left(\frac{d^i g}{d\lambda^i}\right)_{\lambda=0}\,, \quad (i = 0,1,\ldots). \tag{43}$$

If the reference configuration is in the natural state, then from (41), (42) follows that

$$\left.\begin{aligned} Y^{(1)}_{rs,s} &= \gamma F_r\,, \\ \varphi^{(1)}_{rs,s} + e_{rpq} Y^{(1)}_{qp} &= \gamma M_r\,, \end{aligned}\quad (\text{in } C), \right\} \tag{44}$$

$$Y^{(1)}_{rs}\, n_s = f_r\,, \qquad \varphi^{(1)}_{rs}\, n_s = m_r\,, \quad (\text{on } \sigma) \tag{45}$$

(46)
$$\begin{cases} Y^{(2)}_{rs,s} = -2\left(u^{(1)}_{r,l}\, Y^{(1)}_{ls}\right)_{,s} \\ \varphi^{(2)}_{rs,s} + e_{rpq} Y^{(2)}_{qp} = -2\left(u^{(1)}_{r,l}\, \varphi^{(1)}_{ls}\right)_{,s} + e_{rpq}\, \varkappa^{(1)}_{p,l}\left(Y^{(1)}_{lq} - Y^{(1)}_{ql}\right) \end{cases}$$

(47)
$$Y^{(2)}_{rs}\, n_s = -2u^{(1)}_{r,l}\, Y^{(1)}_{ls}\, n_s\,, \quad \varphi^{(2)}_{rs}\, n_s = -2\varkappa^{(1)}_{r,l}\, \varphi^{(1)}_{ls}\, n_s\,.$$

Equations (44), (45), have the same form as in the linear theory and the external loads have to satisfy the well known integral equilibrium equations. At the same time, from equations (46), (47) we shall obtain the compatibility conditions by substituting the terms of the solution of the linear theory in the place of the external loads.

We can show that the necessary conditions of compatibility have the form

(48)
$$\int_C e_{rpq}\, u^{(1)}_{p,l}\left(Y^{(1)}_{ql} + Y^{(1)}_{lq}\right) dC = 0\,.$$

Let $\bar{u}^{(1)}_p$ be a solution of the linear theory: any other solution can then be obtained by adding to $u^{(1)}_p$ an arbitrary infinitesimal rigid displacement admissible by constraints. Thus we can write

(49)
$$u^{(1)}_p = \bar{u}^{(1)}_p + e_{pqs}\, b_q\, y_s\,,$$

where b_q are constants.
Putting

$$a_{rq} = -\int_C \gamma F_r y_q \, dC - \int_\sigma f_r y_q \, d\sigma , \tag{50}$$

and keeping in mind that (44), (45) hold, we shall transform the condition (48) to the form

$$\left[a_{rq} - \delta_{rq} I_1(a)\right] b_q = \frac{1}{2}\left\{ \int_C e_{rpq}\left[\bar{u}_p^{(1)} F_q - \bar{u}_{p,\ell}^{(1)} Y_{\ell q}^{(1)}\right] dC + \int_\sigma e_{rpq} \bar{u}_p^{(1)} f_q \, d\sigma \right\} . \tag{51}$$

In general, relation (51) determines the constants b_q. That means that the rigid displacement which is commonly supposed to be indetermined will be determined if the connection between the linear theory and the non-linear theory is taken into account.
In the special case the determinant $\Delta = \| a_{rq} - I_1(a)\delta_{rq} \|$ may be equal to zero and the equations (51) may admit no solution. In this case the solution of (46), (47) does not exist and the sense of the linear theory is not clear.

In the non-polar case the condition which corresponds to (51) has the same first term, (4), but in the polar case not all external loads are represented in equation (51).

In the case in which Δ is equal to zero, we may not deal with the equilibrium state, admitting the accelerations determined by the expression of the second order in λ .

In this case it would be $\ddot{u}_r^{(1)} = 0$ and $\ddot{u}_r^{(2)} \neq 0$.

We can observe that when the equations which correspond to the derivatives of an arbitrary order with respect to λ are taken into account, then the compatibility conditions have always the form given by (51), the only difference being in the second term. Hence, the condition $\Delta \neq 0$ is in general necessary to obtain the solution of every successive system of equations.

REFERENCES

[1] Cosserat E. and F.: "Sur la mecanique générale" C.R. Acad. Sci. Paris 145, 1139-1142 (1907).

[2] Cosserat E. and F.: "Théorie des corps deformables" Paris, Hermann vi + 266 (1909).

[3] Reissner E.: "Note on the theorem of the symmetry of the stress tensor" Journ. of Mathematics and Physics, vol. XXIII, pp. 192, (1944).

[4] Signorini A.: "Trasformazioni termoelastiche finite" Mem. II, Ann. Matematica pura ed applicata, Ser. IV, 30.1-72 (1949).

[5] Bodászewski S.: "On the asymmetric state of stress and its applications to the mechanics of continuous mediums", Archiwan Mechanicki Stosowanej 5, p. 351 (1953).

[6] Truesdell C. and Toupin R.A.: "The classical field theories" Handbuch der Physik, B.III/I (1960).

[7] Grioli G.: "Elasticità asimmetrica" Ann. Matematica pura ed appl. IV, 4, pp. 389-418 (1960).

[8] Grioli G.: "Onde di discontinuità ed elasticità asimmetrica" Acc. Nazionale del Lincei, S. VIII, vol. XIXX, fasc. 5 (1960).

[9] Aero E.L. and Kuvshinskii E.V.: "Fundamental equations of the theory of elastic media with rotationally interacting particles" Fisika Tverdogo Tela 2, pp. 1399-1409 (1960).

[10] Ericksen J. L., Arch. Rat. Mech. Anal. 4, 231 (1960).

[11] Ericksen J. L., Trans. Soc. Rheol. 4, 29 (1960).

[12] Ericksen J. L., ibidem 5, 23 (1960).

[13] Grioli G.: "Mathematical theory of elastic equilibrium (Recent results)" Ergebnisse der angewandten Mathematik pp. 141-160 (1962).

[14] Toupin R. A.: "Elastic materials with couple stress" Arch. Rat. Mech. and Anal., v. 11 n. 5 (1962).

[15] Mindlin R. D. and Tiersten H. F.: "Effects of couple-stress in linear elasticity", Arch. Rat. Mech. and Anal., v. 11 n. 5 (1962).

[16] Galletto D. "Sulle equazioni in coordinate generali della statica dei continui con caratteristiche di tensione asimmetriche" Ann. dell'Università di Ferrara S. VII, Sci. Mat., vol. X n. 5 (1962).

[17] Grioli G.: "Sulla meccanica dei continui a trasformazioni reversibili con caratteristiche di tensione asimmetriche" Seminari dell'Ist. Nazionale di Alta Matematica, Ed. Cremonese (1962-63).

[18] Bressan A.: "Sui sistemi continui nel caso simmetrico" Ann. Mat. pura ed appl. S. IV, T. 72 (1963).

[19] Bressan A.: "Qualche teorema di cinematica delle deformazioni finite" Atti dell'Ist. Veneto di Scienze, Lettere ed Arti, T. CXXI Cl. di Sc. Mat. e Nat. (1963).

[20] Galletto D.: "Nuove forme per le equazioni in coordinate generali della statica dei continui con caratteristiche di tensione asimmetriche" Ann. d. Scuola Normale Superiore di Pisa S. III, vol. XVII, fasc. IV (1963).

[21] Toupin R.: "Theories of elasticity with couple stress" Arch. Rat. Mech. and Anal. 17, pp. 85-112 (1964).

[22] Mindlin R. D.: "Microstructure in linear elasticity" Arch. Rat. Mech. and Anal., 16, pp. 51-78, (1964).

[23] Grioli G.: "Problemi d'integrazione e formulazione integrale del problema fondamentale dell'elastostatica" Simp. Int. sulle Applic. dell'Analisi alla Fisica Matematica, Cagliari-Sassari, Ed. Cremonese (1964).

[24] Eringen A. C. and Suhubi E.S.: "Nonlinear theory of simple micro-elastic solids" I, Intern. Journ. Eng. Sci., 2, pp. 189-203 (1964).

[25] Suhubi E.S. and Eringen A. C.: "Nonlinear theory of simple microelastic solids" II, Intern. J. Eng. Sci., 2, 389-404 (1964).

[26] Palmov V.A., Prikl. Mat. Mech., 28, 401 (1964).

[27] Galletto D.: "Contributo allo studio dei sistemi continui a trasformazioni reversibili con caratteristiche di tensione asimmetriche" Rend. Sem. Matem. Univ. di Padova (1965).

[28] Schaefer H.: Continui di Cosserat "Lezioni e Conferenze dell'Università di Trieste" Istituto di Meccanica, fasc. 7 (1965).

[29] Galletto D.: "Sistemi incomprimibili a trasformazioni reversibili nel caso asimmetrico" Rendic. Sem. Mat. Univ. Padova (1965-66).

[30] Eringen A. C.: "Linear theory of micropolar elasticity" J. Math. Mech., 15, pp. 909-924 (1966).

[31] Galletto D.: "On continuous media with contact couples" Meccanica n. 3, vol. I (1966).

[32] Sandru N.: "On some problems of the linear theory of asymmetric elasticity" Int. J. Eng. Sci., 81-96 (1966).

[33] Nowacki W.: "Couple stresses in the theory of thermoelasticity" Proc. of IUTAM Symposia, Springer Verlag, Vienna (1966).

[34] Schaefer H.: "Das Cosserat Kontinuum" Zeitschrift f. angewandte Mathematik und Mechanik, B. 47, H. 8, pp. 485-498 (1967).

[35] Agostinelli C.: "Sulla possibilità di sforzi asimmetrici in un corpo elastico elettricamente conduttore in moto vibratorio sotto l'azione di un campo magnetico" Rend. Acc. Naz. Lincei, s. 8, vol. 43 e 44, fasc. 1 (1968).

[36] Nowacki W.: "Propagation of rotation waves in asymmetric elasticity" Bull. de l'Acad. Polonaise d. Sciences, Sér. des sciences techniques, vol. XVI n. 10 (1968).

[37] Nowacki W.: "Green functions for micropolar elasticity" ibidem n. 11-12 (1968).

[38] Nowacki W. and Nowacki W.K.: "Generation of waves in an infinite micropolar elastic solid body" I and II, ibidem vol XVII n. 2 (1969).

[39] Nowacki W. and Nowacki W.K.: "Waves in a micropolar cylinder" I and II, ibidem vol. XVII n. 1, (1969).

[40] Nowacki W. and Nowacki W.K.: "Propagation of monochromatic waves in an infinite micropolar elastic plate", ibidem vol. XVIII, n.1 (1969).

[41] Nowacki W.: "On the completeness of potentials in micropolar elasticity" Archiwum Mechanicki Stosowanej, 2, 21 (1969).

[42] Parfitt V. R. and Eringen A. C.: "Reflection of plane waves from the flat boundary of a micropolar elastic half-space" The Journ. of the Acoustic Society of America, vol. 45 n. 5 (1969).

[43] Eringen A. C.: "Compatibility conditions of the theory of micromorphic elastic solids"Journ. of Math. and Mech., vol. 19, pp. 473-482 (1969).

[44] Grioli G.: "Energia libera e stato tensionale dei continui" Ann. Mat. pura ed applicata s. IV-T, LXXXVII (1970).

[45] Grioli G.: "Microstrutture" I and II, Rend. Acc. Naz. Lincei, s. VIII, vol. XLIX (1970).

[46] Eringen A. C.: "Mechanics of micropolar continua", Pergamon Press (1970).

[47] Kafadar C. B. and Eringen A. C.: "Micropolar media - I The classical theory" Int. J. Engng. Sci., vol. 9, pp. 271-305 (1971).

[48] Kafadar C. B. and Eringen A. C.: "Micropolar media II The relativistic theory" ibidem pp. 307-329 (1971).

R. STOJANOVIC

NONLINEAR MICROPOLAR ELASTICITY

1. Introduction

The Cosserat continuum is a material continuum with "points" which may rotate independently of the displacements. A better interpretation of this continuum is offered through the assignment of rigid triads of vectors to the points, and admitting the triads to rotate independently of the points of the medium.

The pioneers in the generalization of the classical concept of material continua were the brothers Eugène and François Cosserat, who in 1909 developed the theory of a continuum of oriented particles [1] (cf. also [2], [3]). One of the most important contributions made in this theory was the discovery that the stress tensor is not necessarily symmetric, as follows from the second law of Cauchy, and that there is another quantity, to-day called the couple-stress tensor, which characterizes the state of stress in a body together with the stress tensor of Cauchy. Only the antisymmetric part of the Cauchy stress tensor is related to the couple-stress tensor.

The idea of the non-symmetric stress tensor gave rise to numerous generalizations of the classical continuum mechanics, including materials with an arbitrary number of deformable vectors assigned to the points of the medium. Almost all of these generalizations have in the basis definite physical models. In this report we are not going to consider the generalized Cosserat continua, but we restrict our attention to the theory of elasticity of a medium with rigid triads of vectors assigned to its points. These vectors are after Ericksen and Truesdell [4] called the directors.

After the reformulation of the results of Cosserats in the more modern vectorial notation by Sudria [5] in 1935, their work was almost completely ignored till 1958, when Kröner [6] noted that the crystals with dislocations may be regarded as oriented media in the Cosserat sense, and when Günther [7] gave the first tensorial treatment of the statics and kinematics of the Cosserat continuum. In the same year appeared a generalization offered by Ericksen and Truesdell [4].

Günther's considerations were restricted to infinitesimal displacements and rotations. In 1962 Schäfer [8] published a linear, two-dimensional theory of the elastic Cosserat continuum, and Cowin [9] gave the first general, nonlinear theory. Completely independently, without even knowing the work of the Cosserats and other later authors, Aero and Kuvshinskii [10] in 1960 presented a generalization of the classical continuum mechanics, introducing the rotation of the particles into the theory. They obtained the relations already derived by previous authors, but generalizing Hooke's law they developed a linear theory of elasticity for this medium. In their paper they admitted the possibility that particles of a continuum may have rotations independent of the displacements. but they developed a theory with the so-called constrained rotation, which corresponds to what is known to-day as the theory of materials of grade two (Truesdell and Toupin [11], Grioli [12], Toupin [13]). However, in 1963 Kuvshinskii and Aero [14] developed a theory with independent rotations.

Very general approaches to the non-linear theory of elastic Cosserat continua were presented in 1964 by Toupin [15] and in 1965 by Eringen [16]. Toupin derived the basic relations for the Cosserat continua from a very general theory of elastic materials with couple-stresses, and Eringen considered it, under the name of "micropolar elasticity", as a special case of micro-elastic, or micropolar elastic materials, the theory of which was developed a year earlier by Eringen and Suhubi [17]. The micromorphic materials admit three deformable directors.

In all linear theories the strain tensor is not symmetric,

$$\underline{\varepsilon}_{ij} = u_{i,j} - \underline{\epsilon}_{ijk}\underline{\varphi}^{k}.$$

Here u_j is the displacement vector, $\underline{\varphi}^k$ is the angle of rotation of the director triads, and $\underline{\epsilon}_{ijk}$ is the alternating Ricci tensor. Besides the strain tensor $\underline{\varepsilon}_{ij}$ there is also a tensor $\underline{\varkappa}^{k}_{\cdot l}$,

$$\underline{\varkappa}^{k}_{\cdot l} = \underline{\varphi}^{k}_{,l}$$

and the tensors $\underset{\approx}{\varepsilon}$ and $\underset{\approx}{\varkappa}$ are connected with the non-symmetric Cauchy stress tensor t^{ij} and with the couple stress tensor $m_k^{\cdot l}$.

In the non-linear theories considered, is an energy function w and it is assumed that it is a function of the deformation gradients, of the components $d^k_{(\underline{\alpha})}$ of the directors, and of the material gradients, $d^k_{(\underline{\alpha});\underline{k}}$ of the directors. When linearized, the deformation tensors of the non-linear theories reduce to $\underline{\varepsilon}$ and $\underline{\varkappa}$.

However, in the theory of the generalized Cosserat continua it is demonstrated that the energy function can not be a function of the components of the directors $d^k_{(\underline{\alpha})}$ but only of the gradients (cf. Stojanovic and Djuric [18] , Stojanovic [19]). As a consequence, in the theory of elastic materials with rigid director triads after linearization the state of strain is determined by the <u>symmetric</u> strain tensor $e_{ij} = \underline{\varepsilon}_{ij}$ and by the tensor $\underset{\approx}{\varkappa}$.

In what follows the non-linear theory of the

elastic Cosserat continuum will be developed assuming that the internal energy function is a function of the components of the directors. After proving that this assumption is inadmissible, the constitutive relations for stress and couple-stress will be derived. It appears that in the linear approximation the antisymmetric part of the stress tensor vanishes.

2. The Cosserat Continuum

We consider a continuum medium to each point of which are assigned triads of vectors $d_{(\alpha)}$, $\alpha = 1,2,3$ called the directors. The triads of directors may rotate independently of the motions of the points, but during the motion the components $d^k_{(\alpha)}$ of the directors have to satisfy six equations of constraints,

$$\underset{\sim}{d}_{(\alpha)} \cdot \underset{\sim}{d}_{(\beta)} = \underline{d}_{\alpha\beta} = \text{const.} \tag{2.1}$$

and at each point of the body there are six degrees of freedom, three deformations of position and three deformations of orientation.

At an initial moment t_0 of time let the points of the body be referred to a system of (material) coordinates $\underline{X}^K$, and let the components of the directors at this configuration be denoted by $\underline{D}^K_{(\alpha)}$. In general we have $\underset{\sim}{D}_{(\alpha)} = \underset{\sim}{D}_{(\alpha)}(\underline{X}^1, \underline{X}^2, \underline{X}^3)$, but without loss of generality we may assume that the directors in the initial configuration represent fields of parallel vectors, i.e.

$$\partial_L \underset{\sim}{D}_{(\alpha)} = D^K_{(\alpha),L} \underset{\sim}{G}_K = 0 \quad , \tag{2.2}$$

where ∂_L stands for the partial derivation, $\underset{\sim}{G}_K$ are the base vectors for the coordinate system X^K, and ",L" denotes the (partial) covariant differentiation (1). At an arbitrary moment t of time the points $\underset{\sim}{X}$ of the body occupy the positions $\underset{\sim}{x}$ in the space, and x^k are spatial coordinates of points of the body. The equations of motion are

$$\begin{aligned} x^k &= x^k(X^1, X^2, X^3; t) \\ d^k_{(\alpha)} &= d^k_{(\alpha)}(X^1, X^2, X^3; t) \end{aligned} \tag{2.3}$$

where the equations of motion for the directors have to be compatible with the constraints (2.1). Let us denote the base vectors for spatial coordinates by g_k and the fundamental tensor corresponding to these coordinates by g_{kl}.
We assume that

$$\det D^K_{(\alpha)} \neq 0 \quad , \quad \det d^k_{(\alpha)} \neq 0 \tag{2.4}$$

---

(1) We adopt the notation of the theory of two-point tensor fields, with " ;L" denoting the total covariant derivatives, e.g.

$$(\ldots)_{;L} = (\ldots)_{,L} + (\ldots)_{,l}\, x^l_{;L} \quad ,$$

$$(\ldots)_{;l} = (\ldots)_{,l} + (\ldots)_{,L}\, X^L_{;l} \quad .$$

at all points of the body. The components $D_K^{(\alpha)}$, $d_K^{(\alpha)}$ of the reciprocal triads $\underset{\sim}{D}^{(\alpha)}$, $\underset{\sim}{d}^{(\alpha)}$ of the directors satisfy the relations

$$D_K^{(\alpha)} D_{(\beta)}^K = \delta_\beta^\alpha \quad , \quad D_K^{(\alpha)} D_{(\alpha)}^L = \delta_K^L \ , \qquad (2.5)$$
$$d_k^{(\alpha)} d_{(\beta)}^k = \delta_\beta^\alpha \quad , \quad d_k^{(\alpha)} d_{(\alpha)}^l = \delta_k^l \ .$$

From (2.1) and (2.5) follow the equations of constraints for the rates of the directors, (2)

$$d_{(i}^{(\alpha)} \dot{d}_{(\alpha)j)} = 0 \ , \qquad (2.6)$$

where the superposed dot denotes the material time derivative. Similarly, if we consider two possible configurations of the directors, the differences of their components, i.e. the variations $\delta d_{(\alpha)j}$ have to satisfy the constraints

$$d_{(i}^{(\alpha)} \delta d_{(\alpha)j)} = 0 \ . \qquad (2.7)$$

If we introduce the gyration tensor $\underset{\sim}{\nu}$,

$$d_i^{(\alpha)} \dot{d}_{(\alpha)j} = \nu_{ij} = -\nu_{ji} \ , \qquad (2.8)$$

---

(2) The round brackets denote the symmetric part of quantity considered with respect to the indices involved in the brackets, and the square brackets denote the alternation of the indices involved, i.e. the formation of the antisymmetric part. Ordinal numbers (greek indices in the round brackets) are excluded from these operations. E.g.

$$d_{(i}^{(\alpha)} \dot{d}_{(\alpha)j)} = \frac{1}{2} \left(d_i^{(\alpha)} \dot{d}_{(\alpha)j} + d_j^{(\alpha)} \dot{d}_{(\alpha)i} \right).$$

according to (2,6) this tensor has only three independent components, and we have

$$\dot{d}_{(\alpha)j} = \nu_{ij} d^i_{(\alpha)} . \tag{2.9}$$

It should be noted that the director velocities $\dot{d}_{(\alpha)i}$ are composed of two parts. One part is the velocity induced by the motion of the bearing points. If $w_{ij} = v_{[j,i]}$ is the vorticity at the points of the body, with $v^i = \dot{x}^i$, the induced director velocity is $\dot{\underset{I}{d}}_{(\alpha)i} = w_{mi} d^m_{(\alpha)}$. The second part is a relative velocity $\dot{\underset{R}{d}}_{(\alpha)i}$ which is completely independent of the motion of the points of the medium.

Since the directors represent rigid triads, an orthogonal tensor $\underset{\sim}{\chi} = \underset{\sim}{\chi}(\underset{\sim}{X}, t)$ may be introduced such that

$$d^k_{(\alpha)} = \chi^k_{\cdot K} D^K_{(\alpha)} , \quad D^K_{(\alpha)} = \chi^K_{\cdot l} d^l_{(\alpha)} , \tag{2.10}$$

and

$$\dot{d}^k_{(\alpha)} = \dot{\chi}^k_{\cdot K} \chi^K_{\cdot l} d^l_{(\alpha)} \tag{2.11}$$

and from this and (2.9) it follows that

$$\dot{\chi}^k_{\cdot K} \chi^K_{\cdot l} = -\nu^k_{\cdot l} . \tag{2.12}$$

Various physical models which in the approximation reduce to the Cosserat continuum yield the same expressions for the momentum, moment of momentum and kinetic energy. Let ϑ be an arbitrary portion of the body considered and let ϱ be the (macroscopic) mass density of the body. The momentum is

$$\underset{\sim}{K} = \int_{\vartheta} \varrho \underset{\sim}{v} \, d\vartheta \tag{2.13}$$

where $\underset{\sim}{v} = \dot{\underset{\sim}{r}} = \dot{x}^k \underset{\sim}{g}_k$ is the velocity vector for the points of the body and $\underset{\sim}{r}$ is the position vector. The moment of momentum is given by the expression

$$\underset{\sim}{L} = \int_{\vartheta} \varrho (\underset{\sim}{r} \times \underset{\sim}{v} + I^{\alpha\beta} \underset{\sim}{d}_{(\alpha)} \times \dot{\underset{\sim}{d}}_{(\beta)}) \, d\vartheta . \tag{2.14}$$

Here "x" denotes the vectorial product, and $I^{\alpha\beta}$ are coefficients of the microinertia. E. g. if the body is composed of rigid particles of microdensity ϱ' and volume ϑ', and if ξ' are local Cartesian coordinates with origins at the centres of gravity of the particles,

$$\varrho I^{\alpha\beta} \vartheta' = \int_{\vartheta'} \varrho' \xi^{\alpha} \xi^{\beta} \, d\vartheta' .$$

The kinetic energy of the portion ϑ of the body

$$T = \frac{1}{2} \int_{\vartheta} \varrho (\underset{\sim}{v} \cdot \underset{\sim}{v} + I^{\alpha\beta} \dot{\underset{\sim}{d}}_{(\alpha)} \cdot \dot{\underset{\sim}{d}}_{(\beta)}) \, d\vartheta . \tag{2.15}$$

The tensor

$$I^{\alpha\beta} \underset{\sim}{d}_{(\alpha)} \times \dot{\underset{\sim}{d}}_{(\beta)} \equiv \underset{\sim}{\sigma} , \tag{2.16}$$

in (2.14) represents the microspin and it is an antisymmetric tensor with the components

$$(2.17)\quad \sigma^{ij} = I^{\alpha\beta} d^{[i}_{(\alpha)} \dot{d}^{j]}_{(\beta)} = (i^{im} \nu_m^{\cdot j})_{[ij]} \,, \quad (i^{im} \equiv I^{\alpha\beta} d^i_{(\alpha)} d^m_{(\beta)}) \,.$$

Using the constraints (2.6) and (2.17) we easily find that the part of kinetic energy which corresponds to the motion of the director triads may be represented also in terms of the spin,

$$(2.18)\qquad I^{\alpha\beta} \dot{\underset{\sim}{d}}_{(\alpha)} \cdot \dot{\underset{\sim}{d}}_{(\beta)} = \sigma^{ij} \nu_{ij} \,.$$

The Cosserat continuum has six degrees of freedom, three motions $x^i(t)$ of its points and three rotations at the points. Kafadar and Eringen [20] attempted ro represent the rotations in terms of three independent angles, but the angles are not objective quantities and the theory based on this representation looses in generality. We prefer to consider the directors here as the basic entities. Since there are six constraints satisfied by the components of the directors, there are in fact three rotational degrees of freedom for the director triads.

To derive the equations of motion and the constitutive relations we have to introduce certain assumptions. The assumptions we are going to introduce are the assumptions usually accepted in the continuum mechanics and in the theory of elasticity, but generalized in order to introduce the influence of orientation.

1. We assume that the medium considered is

superelastic, such that there exists an internal energy function w (per unit mass) of the form

$$w = w\,(x^k_{;K}\,,\, d^l_{(\alpha)},\, d^l_{(\alpha);K}\,,\, \eta\,,\, X^K)\;, \qquad (2.19)$$

where η is the specific entropy per unit mass of the body, $x^k_{;K}$ are the deformation gradients and $d^l_{(\alpha);K}$ are the director gradients.

2. We assume that w is an objective function, independent of any uniform rotation of the spatial coordinate system. This assumption yields the system of three partial differential equations (cf. Toupin [21])

$$\left[g^{il}\left(\frac{\partial w}{\partial x^l_{;L}}\, x^j_{;L} + \frac{\partial w}{\partial d^l_{(\alpha)}}\, d^j_{(\alpha)} + \frac{\partial w}{\partial d^l_{(\alpha);K}}\, d^j_{(\alpha);K} \right)\right]_{[ij]} = 0\,. \qquad (2.20)$$

3. We assume that the conservation of mass law holds,

$$\frac{d}{dt}\,(\rho\, d\vartheta) = 0\;. \qquad (2.21)$$

4. We assume the principle of virtual work in the form

$$\int_\vartheta \rho(\dot v^i \delta x_i + I^{\alpha\beta}\ddot d^{\,i}_{(\beta)}\delta d_{(\alpha)i})\,d\vartheta = \int_\vartheta \rho(f^i\delta x_i + l^{(\alpha)i}\delta d_{(\alpha)i})\,d\vartheta + \oint_a (T^i\delta x_i + H^{(\alpha)i}\delta d_{(\alpha)i})\,da\;, \qquad (2.22)$$

where f^i and $l^{(\alpha)i}$ are certain assigned body forces and T^i and $H^{(\alpha)i}$ certain assigned surface tractions at the bounding surface a of ϑ. Since the vectors $\delta d_{(\alpha)i}$ are not independent, simultaneously with (2.22) we have to consider the constraints (2.7).

5. We finally assume that the equations of motion and the constitutive relations have to be compatible with the laws of thermodynamics. The first law we assume in the form of the energy balance equation,

$$\frac{d}{dt}(T + W) = A + Q \tag{2.23}$$

where W is the total internal energy of the portion ϑ of the body considered, A is the rate of mechanical working and Q is the rate of the non-mechanical production of energy. The second law of thermodynamics will be assumed in the form of the local Clausius-Duhem inequality,

$$\rho\theta\dot{\eta} + q^j_{,j} + \rho h + \frac{1}{\theta}\theta_{,k}q^k \geq 0 , \tag{2.24}$$

where θ is the temperature, q^j is the heat flux vector and h is the heat production per unit mass, such that

$$Q = \oint_a q^k da_k + \int_\vartheta \rho h \, d\vartheta . \tag{2.25}$$

3. The Principle of Virtual Work and the Equations of Motion

To derive the equations of motion from the principle of virtual work (2.22) we shall apply the theorem of Piola in the form presented by Truesdell and Toupin [11], but modifying it for oriented media. The original theorem, in the non-polar case, asserts that the condition (2.22) for virtual translations is equivalent to Cauchy's first law of motion, and for rigid virtual displacements, to Cauchy's second law. In this statement a virtual translation is a field $\delta \underset{\sim}{r}$ such that $\delta x_{i,j} = 0$, and a rigid virtual displacement is a field $\delta \underset{\sim}{r}$ such that $\delta x_{(i,j)} = 0$.

In the case of Cosserat continua we consider not only the virtual displacements $\delta \underset{\sim}{r}$ of the points of the medium, but also the virtual changes $\delta \underset{\sim}{d}_{(\alpha)}$ of the fields of the directors. For virtual translations we assume that the directors remain unchanged, and (2.22) reduces to

$$\int_{\vartheta} \rho \dot{v}^i \delta x_i \, d\vartheta = \int_{\vartheta} \rho f^i \delta x_i \, d\vartheta + \oint_{a} T^i \delta x_i \, da \,, \qquad (3.1)$$

and

$$\delta x_{i,j} = 0 \qquad (3.2)$$

represent the constraints to be satisfied by the virtual displacements. Introducing the Lagrangean multipliers $-t^{ij}$ and adding to (3.1) the relation

(3.3) $$0 = -\int t^{ij}\delta x_{i,j}\, d\vartheta = -\oint_a t^{ij} n_j \delta x_i da + \int_\vartheta t^{ij}_{,j}\delta x_i\, d\vartheta \ ,$$

we obtain an expression equivalent to (3.1) for arbitrary variations δx_i. By the usual procedure we obtain now the equations of motion

(3.4) $$\rho \dot{v}^i = \rho f^i + t^{ij}_{,j} \ ,$$

and the boundary conditions

(3.5) $$T^i - t^{ij} n_j = 0$$

where n_j is the unit normal to the bounding surface a.

The field of rigid virtual displacements may be represented by

(3.6) $$\delta \underset{\sim}{r} = \underset{\sim}{\omega} \times \underset{\sim}{r} + \underset{\sim}{a} \ ,$$

where $\underset{\sim}{\omega}$ is a virtual rotation and $\underset{\sim}{a}$ a virtual translation, and both $\underset{\sim}{\omega}$ and $\underset{\sim}{a}$ are arbitrary constant vectors. The translation is already discussed and we assume now that $\underset{\sim}{a} = 0$. The rotation $\underset{\sim}{\omega}$ induces also the rotations of the directors and assuming that there are no independent variations of the directors, we have

(3.7) $$\delta \underset{\sim}{d}_{(\alpha)} = \underset{\sim}{\omega} \times \underset{\sim}{d}_{(\alpha)} \ .$$

Elimination of $\underset{\sim}{\omega}$ from (3.6) and (3.7) yields the constraints

(3.8a) $$\delta x_{(i,j)} = 0 \ ,$$

$$\delta d_{(\alpha)j,i} + d^{m}_{(\alpha),i}\,\delta x_{m,j} = 0 \,. \tag{3.8b}$$

Introducing now the Lagrangean multipliers $-t^{(ij)}$ and $-h^{(\alpha)ji}$ with the constraints (3.8) we obtain a relation equivalent to (2.22),

$$\begin{aligned}\int_{\vartheta}\varrho(\dot{v}^{i}\delta x_{i} + I^{\alpha\beta}\ddot{d}^{j}_{(\beta)}\delta d_{(\alpha)j})\,d\vartheta = \int_{\vartheta}[\varrho f^{i}\delta x_{i} + \varrho l^{(\alpha)j}\delta d_{(\alpha)j} +\\ -(t^{(ij)} + d^{[i}_{(\alpha),k}h^{(\alpha)j]k})\,\delta x_{i,j} - h^{(\alpha)jk}\delta d_{(\alpha)j,k}]\,d\vartheta +\\ + \oint_{a}(T^{i}\delta x_{i} + H^{(\alpha)j}\delta d_{(\alpha)j})\,da\end{aligned} \tag{3.9}$$

for arbitrary variations δx_{i}, $\delta d_{(\alpha)j}$. In (3.9), in view of $(3.8)_{1}$, we have put

$$h^{(\alpha)jk}d^{m}_{(\alpha),k}\,\delta x_{m,j} = d^{[m}_{(\alpha),k}h^{(\alpha)j]k}\,\delta x_{m,j}\,. \tag{3.10}$$

After the application of the divergence theorem to the volume integral on the right-hand side of (3.9), this becomes

$$\begin{aligned}\int_{\vartheta}\varrho(\dot{v}^{i}\delta x_{i} + I^{\alpha\beta}\ddot{d}^{j}_{(\beta)}\delta d_{(\alpha)j})\,d\vartheta = \int_{\vartheta}\{[\varrho f^{i} + t^{(ij)}_{,j} + (d^{[i}_{(\alpha),k}h^{(\alpha)j]k})_{,j}]\delta x_{i} +\\ + [\varrho l^{(\alpha)j} + h^{(\alpha)jk}_{,k}]\,\delta d_{(\alpha)j}\}\,d\vartheta + \oint_{a}[(T^{i} - t^{(ij)}n_{j} +\\ - d^{[i}_{(\alpha),k}h^{(\alpha)j]k}n_{j})\,\delta x_{i} + (H^{(\alpha)j} - h^{(\alpha)jk}n_{k})\,\delta d_{(\alpha)j}]\,da\,.\end{aligned} \tag{3.11}$$

However, the directors form rigid triads and the variations have to satisfy the constraints (2.7). If we introduce three independent variations $\delta\varphi_{ij}=-\delta\varphi_{ji}$, such that

(3.12) $$\delta d_{(\alpha)j} = \delta\varphi_{mj} d^m_{(\alpha)} ,$$

the equations (2.7) will be identically satisfied. For arbitrary δx_i and $\delta\varphi_{mj}$, (3.11) will be satisfied if the following relations are valid in the body

(3.13) $$\rho\dot{v}^i = \rho f^i + (t^{(ij)} + d^{[i}_{(\alpha),k} h^{(\alpha)j]k})_{,j} ,$$

(3.14) $$\rho\dot{\sigma}^{ij} = \rho d^{[i}_{(\alpha)} l^{(\alpha)j]} + d^{[i}_{(\alpha)} h^{(\alpha)j]k}_{,k} ,$$

where σ^{ij} is given by (2.17), and on the bounding surface

(3.15) $$0 = T^i - (t^{(ij)} + d^{[i}_{(\alpha),k} h^{(\alpha)j]k}) n_j ,$$

(3.16) $$0 = d^{[i}_{(\alpha)} H^{(\alpha)j]} - d^{[i}_{(\alpha)} h^{(\alpha)j]k} n_k .$$

Comparing (3.13) and (3.15) with (3.4) and (3.5) we see that the antisymmetric part of the <u>stress tensor</u> t^{ij} is related to the <u>director stresses</u> $h^{(\alpha)ij}$ by

(3.17) $$t^{[ij]} = d^{[i}_{(\alpha),k} h^{(\alpha)j]k} .$$

Introducing further the notation

(3.18) $$d^{[i}_{(\alpha)} h^{(\alpha)j]k} \equiv m^{ijk} ,$$

$$d^{[i}_{(\alpha)} l^{(\alpha)j]} = l^{ij} \qquad (3.19)$$

$$d^{[i}_{(\alpha)} H^{(\alpha)j]} = H^{ij} \qquad (3.20)$$

where m^{ijk} is the couple-stress tensor, and l^{ij} and H^{ij} are certain body and surface couples, besides the equations of motion (3.4) and the boundary conditions we now have an additional set of three equations of motion

$$\rho \dot{\sigma}^{ij} = \rho l^{ij} - t^{[ij]} + m^{ijk}_{\ ,k} \quad , \qquad (3.21)$$

and the corresponding boundary conditions for the couple-stress

$$H^{ij} = m^{ijk} n_k \quad . \qquad (3.22)$$

The equations (3.4) and (3.21) represent the complete set of the equations of motion of a Cosserat continuum, and (3.5) and (3.22) are the boundary conditions. The state of stress in this continuum is determined in terms of the non-symmetric Cauchy stress $\underset{\sim}{t}$ and the couple-stress $\underset{\sim}{m}$. These two stress tensors are determined through the constitutive relations.

4. The Constitutive Relations

In the case of an elastic medium the constitutive relations may be derived from the principles of thermodynamics. According to (2.23) the first law of thermodynamics may be written in the form

$$\frac{d}{dt}\int_{\vartheta}\rho\left(\frac{1}{2}v^i v_i + \frac{1}{2}I^{\alpha\beta}\dot{d}_{(\alpha)j}\dot{d}^j_{(\beta)} + w\right)d\vartheta = \int_{\vartheta}\rho(f^i v_i + \tag{4.1}$$

$$+ l^{(\alpha)j}\dot{d}_{(\alpha)j} + h)d\vartheta + \oint_a (t^{ik}v_i + h^{(\alpha)jk}\dot{d}_{(\alpha)j} + q^k)da_k$$

Using (2.9), (2.18), (3.4) and (3.14), performing the differentiation on the left-hand side of (4.1) and applying the divergence theorem to the surface integral, from (4.1) follows the reduced energy equation in the local form,

$$\rho\dot{w} = t^{ij}v_{i,j} + h^{(\alpha)jk}d_{(\alpha)j,k} + \rho h + q^k_{,k} . \tag{4.2}$$

Combining now (4.2) with the local Clausius-Duhem inequality (2.25) we obtain an inequality

$$\rho\dot{w} - t^{ij}v_{i,j} - h^{(\alpha)jk}d_{(\alpha)j,k} + \rho\theta\dot{\eta} + \frac{1}{\theta}\theta_{,k}q^k \geq 0 \tag{4.3}$$

which has to be satisfied by any admissible process.

Writing

$$v_{i,j} = g_{il}X^K_{;j}\dot{x}^l_{;K}$$

and using (2.19), the inequality (4.3) may be written in the form

$$\left(\rho g^{il}\frac{\partial w}{\partial x^l_{;K}} - t^{ij}X^K_{;j}\right)\dot{x}^l_{;K} + \rho\frac{\partial w}{\partial d^l_{(\alpha)}}\dot{d}^l_{(\alpha)} +$$

$$+\left(\rho g^{jl}\frac{\partial w}{\partial d^l_{(\alpha);K}} - h^{(\alpha)jk}X^K_{;k}\right)\dot{d}_{(\alpha)j;K} + \rho\left(\frac{\partial w}{\partial\eta} + \theta\right)\dot{\eta} + \tag{4.4}$$

$$+ \frac{1}{\theta}\theta_{,k}q^k \geq 0 .$$

In the theory of generalized Cosserat continua, i.e. in the theory of continua with deformable directors, the inequality (4.4) will be satisfied for arbitrary rates $\dot{x}^l_{;K}$, $\dot{d}^l_{(\alpha)}$, $\dot{d}_{(\alpha)j;K}$, $\dot{\eta}$ only if the multipliers with these rates vanish. This implies that the internal energy function does not depend explicitly on the components of the directors,

$$\frac{\partial w}{\partial d^l_{(\alpha)}} = 0 \quad , \tag{4.5}$$

and also

$$h^{(\alpha)jk} = \rho g^{jl} \frac{\partial w}{\partial d^l_{(\alpha);K}} x^k_{;K} \quad . \tag{4.6}$$

In the special case, when the directors are only free to rotate as rigid triads, these implications of the general theory should not be violated and the constitutive relations have to be compatible with (4.5) and (4.6). Hence, assuming (4.5) and replacing the rates $\dot{d}_{(\alpha)j}$ from (2.9), we see that the inequality (4.4) will be satisfied for arbitrary rates $\dot{x}^l_{;K}$ and $\dot{\eta}$ and for arbitrary gyrations ν_{ij} and gyration gradients $\nu_{ij;K}$ only if the following relations are valid during the motion:

$$t^{ij} = \rho g^{il} \frac{\partial w}{\partial x^l_{;K}} x^j_{;K} \quad , \tag{4.7}$$

$$d^{[i}_{(\alpha)} h^{(\alpha)j]k} = \rho \left(d^i_{(\alpha)} g^{jl} \frac{\partial w}{\partial d^l_{(\alpha);K}} x^k_{;K} \right)_{[ij]} \quad , \tag{4.8}$$

$$(4.9)\qquad \theta = -\frac{\partial w}{\partial \eta}$$

$$(4.10)\qquad d^{[i}_{(\alpha),k}h^{(\alpha)j]k} = \rho\left(d^{i}_{(\alpha),k}g^{jl}\frac{\partial w}{\partial d^{l}_{(\alpha);K}}x^{k}_{;K}\right)_{[ij]},$$

and the inequality reduces to

$$(4.11)\qquad \frac{1}{\theta}\theta_{,k}q^{k} \geq 0 .$$

The equations (4.7) represent the constitutive relations for the antisymmetric stress tensor t^{ij}, and according to (3.18), the constitutive relations for the couple-stress tensor are given by (4.8).

From (3.17) we see that the left-hand side of (4.10) is equal to the antisymmetric part of the stress tensor. Combining it with the antisymmetric part of (4.7) we obtain

$$(4.12)\qquad \left[g^{il}\left(\frac{\partial w}{\partial x^{l}_{;K}}x^{j}_{;K} + \frac{\partial w}{\partial d^{l}_{(\alpha);K}}d^{j}_{(\alpha);K}\right)\right] = 0 .$$

This relation, however, coincides with (2.20). Thus, (4.10) does not represent an independent set of constitutive relations, but coincides with the requirement that w is an objective function.

From (2.1) it follows that the director gradients have to satisfy the relations

$$(4.13)\qquad d^{(\alpha)}_{i}d_{(\alpha)j;K} + d^{(\alpha)}_{j}d_{(\alpha)i;K} = 0$$

Introducing an antisymmetric tensor $\Phi_{ijK} = -\Phi_{jiK}$ we may represent the gradients $d_{(\alpha)j;k}$ by

$$d_{(\alpha)j;K} = \Phi_{ijK} d^i_{(\alpha)} \, , \tag{4.14}$$

and the relations (4.13) will identically be satisfied. The tensor Φ_{ijK} has only nine independent components and owing to its antisymmetry the dual tensor $\Phi^k_{\cdot K}$ may be introduced such that

$$\Phi^k_{\cdot K} = \frac{1}{2} \epsilon^{kij} \Phi_{ijK} \; ; \; \Phi_{ijK} = \epsilon_{ijk} \Phi^k_{\cdot K} \; , \tag{4.15}$$

where ϵ^{kij} is the Ricci's alternating tensor. The connection between the tensor $\underset{\sim}{\Phi}$ and the orthogonal tensor $\underset{\sim}{\chi}$ may be easily established using (2.10),

$$\Phi_{ijK} = \chi^L_{\cdot i} \chi_{jL;K} \; , \tag{4.16}$$

with

$$\chi_{jL} = g_{jm} \chi^m_{\cdot L} \, . \tag{4.17}$$

The constitutive equations (4.8) for the couple-stress become now

$$m^{ijk} = \rho \frac{\partial w}{\partial \Phi_{ijK}} x^k_{;K} \; , \tag{4.18}$$

where we assume

$$\frac{\partial w}{\partial \Phi_{ijK}} = - \frac{\partial w}{\partial \Phi_{jiK}} \, . \tag{4.19}$$

For the couple-stress tensor m^{ijk} we introduce its dual $m_l^{\cdot k}$,

$$\begin{aligned} m_l^{\cdot k} &= \frac{1}{2} \epsilon_{lij} m^{ijk} \, , \\ m^{ijk} &= \epsilon^{ijl} m_l^{\cdot k} \, , \end{aligned} \tag{4.20}$$

and (4.16) gives

(4.21) $$m_i^{\cdot k} = \frac{1}{2}\rho \frac{\partial w}{\partial \Phi^l_{\cdot K}} x^k_{;K} .$$

The internal energy function may be considered now as a function of 19 independent variables $x^l_{;K}$, $\Phi^l_{\cdot K}$ and η and of the material coordinates X^K. However, w is not an arbitrary function of the variables, but has to satisfy the objectivity condition (4.12) which reduces now to

(4.22) $$\left[g^{il}\left(\frac{\partial w}{\partial x^l_{;K}} x^j_{;K} + \frac{\partial w}{\partial \Phi^l_{\cdot K}} \Phi^j_{\cdot K}\right)\right]_{[ij]} = 0 .$$

These conditions represent a set of three linear partial differential equations and the internal energy function is an arbitrary function of the integrals of (4.22). The system admits 22-3=19 independent integrals. The integrals are the material tensors

(4.23) $$C_{KL} = C_{LK} = g_{mn} x^m_{;K} x^n_{;L} ,$$
$$F_{KL} = g_{mn} x^m_{;K} \Phi^n_{\cdot L} ,$$

and the variables η and X^K and we have

(4.24) $$w = w(C_{KL}, F_{KL}, \eta, X^K)$$

The expressions for the stress tensor and for the dual couple-stress tensor finally read

$$t^{ij} = \varrho\left(2\frac{\partial w}{\partial C_{KL}}x^i_{;K}x^j_{;L} + \frac{\partial w}{\partial F_{KL}}\Phi^i_{\cdot L}x^j_{;K}\right), \qquad (4.25)$$

$$m_l^{\cdot k} = \frac{1}{2}\varrho g_{lm}\frac{\partial w}{\partial F_{KL}}x^m_{;K}x^k_{;L}. \qquad (4.26)$$

5. Linearization, Isotropic Media

Let the points of an elastic Cosserat medium suffer infinitesimal displacements u_i with infinitesimal displacement gradients and let the director triads suffer infinitesimal rotations. In this case we may write

$$x^k = X^K\delta^k_K + u^k, \qquad (5.1)$$

and the orthogonal tensor $\underset{\sim}{\chi}$ may be represented in the form

$$\chi^k_{\cdot K} = \delta^k_K + \varphi^{\cdot k}_K, \qquad (5.2)$$

where $\varphi^{\cdot k}_K = -\varphi^k_{\cdot K}$ is an infinitesimal rotation. According to (4.16), the components of the tensor $\underset{\sim}{\Phi}$ will be

$$\Phi_{ijK} = \varphi_{ij,k}\delta^k_K, \qquad (5.3)$$

and if φ^l is the vectorial representation of the angle of rotation,

$$\varphi^l = \frac{1}{2}\epsilon^{lij}\varphi_{ij}, \qquad (5.4)$$

we have

$$\Phi^{l}_{\cdot K} = \varphi^{l}_{,k}\,\delta^{k}_{K} . \tag{5.5}$$

From (4.23) and (5.1) we easily find in the linear approximation

$$F_{KL} = \varphi_{k,l}\,\delta^{k}_{K}\delta^{l}_{L} = \varkappa_{kl}\,\delta^{k}_{K}\delta^{l}_{L} . \tag{5.6}$$

For the deformation of position it is convenient to introduce the strain tensor $\underset{\sim}{E}$,

$$E_{KL} = -\frac{1}{2}(G_{KL} - C_{KL}) \tag{5.7}$$

and the spatial strain tensor e_{kl} ,

$$e_{kl} = E_{KL}X^{K}_{;k}X^{L}_{;l} , \tag{5.8}$$

which for infinitesimal deformations reduces to

$$e_{kl} \approx E_{KL}\,\delta^{K}_{k}\delta^{L}_{l} \approx u_{(k,l)} . \tag{5.9}$$

If we further assume that the material in the initial configuration is homogeneous with density ρ_0 and if we write w for $\rho_0 w$, the constitutive relations (4.25) and (4.26) become now

$$t^{ij} = \frac{\partial w}{\partial e_{ij}} + \frac{\partial w}{\partial \varkappa_{jl}}\varkappa^{i}_{\cdot l} \tag{5.10}$$

$$m_{l}^{\cdot k} = \frac{1}{2} g_{lm}\frac{\partial w}{\partial \varkappa_{mk}} . \tag{5.11}$$

The function $\mathfrak{w}$ may be approximated by a polynomial in the components of the tensors $\underset{\sim}{e}$ and $\underset{\sim}{\varkappa}$. If there are no initial stresses, in the linear theory of elasticity $\mathfrak{w}$ is a quadratic polynomial,

$$\mathfrak{w} = A^{ijkl} e_{ij} e_{kl} + B^{ijkl} \varkappa_{ij} \varkappa_{kl} + C^{ijkl} e_{ij} \varkappa_{kl} \tag{5.12}$$

where $\underset{\sim}{A}$, $\underset{\sim}{B}$ and $\underset{\sim}{C}$ are tensorial coefficients independent of e and $\underset{\sim}{\varkappa}$.

The second term on the right-hand side of (5.10) is of second order and in the linear theory the antisymmetric part of the stress tensor vanishes. To the same conclusion we may come from (3.17), which gives

$$t^{[ij]} = (\varkappa^{i}_{\cdot k} m^{jk})_{[ij]} \; . \tag{5.13}$$

The antisymmetric part of the stress tensor represents thus an effect of the second order.

For isotropic materials the tensors $\underset{\sim}{A}$ and $\underset{\sim}{B}$ in (5.12) are of the form

$$\begin{aligned} A^{ijkl} &= \alpha_1 g^{ij} g^{kl} + \alpha_2 g^{ik} g^{jl} + \alpha_3 g^{il} g^{jk} \; , \\ B^{ijkl} &= \beta_1 g^{ij} g^{kl} + \beta_2 g^{ik} g^{jl} + \beta_3 g^{il} g^{jk} \; , \end{aligned} \tag{5.14}$$

where the coefficients $\alpha_1, \ldots, \beta_3$ are constants. The tensor C^{ijkl} vanishes since $\underset{\sim}{\varkappa}$ is an axial tensor. For the symmetric part of the stress tensor we obtain in the linear approximation

Hooke's law, and for the couple-stress tensor the linear constitutive relation reads

(5.15) $$m_i^{\cdot k} = \beta_1 \varkappa^{\cdot i}_{\cdot i} \delta^k_i + \beta_2 \varkappa_i^{\cdot k} + \beta_3 \varkappa^k_{\cdot i} \quad .$$

6. Constrained Rotations

Let us finally make a remark on the case when the rotations of the directors coincide with the rotations (vorticity) at the points of the medium, i.e.

(6.1) $$\nu_{ij} = w_{ij} = v_{[j,i]}$$

In this case we find from (2.9)

(6.2) $$\dot{d}_{(\alpha)j} = w_{ij} d^i_{(\alpha)}$$

and the energy equation (4.2) becomes

(6.3) $$\rho \dot{w} = t^{ij} v_{i,j} + d^i_{(\alpha)} h^{(\alpha)jk} w_{ij,k} + d^i_{(\alpha),k} h^{(\alpha)jk} w_{ij} + \rho h + q^k_{,k} \;,$$

or, using (3.17) and (3.18),

(6.4) $$\rho \dot{w} = t^{(ij)} d_{ij} + m^{ijk} w_{ij,k} + \rho h + q^k_{,k} \;,$$

where $d_{ij} = v_{(i,j)}$ is the rate of strain tensor. (6.4) has the same form as the energy equation in the theory of the elastic materials of grade two in which

$$w = w(x^k_{;K}, x^k_{;KL}, \eta, X^k) \quad , \tag{6.5}$$

(cf. Truesdell and Toupin [11], Toupin [13, 15]).

REFERENCES

[1] E. and F. Cosserat: "Sur la mécanique générale", C.R. Acad. Sc. Paris, 145, 1139-1142 (1907).

[2] E. and F. Cosserat: "Sur la théorie des corps minces", C.R. Acad. Sci. Paris, 146, 169-172 (1908).

[3] E. and F. Cosserat: "La théorie des corps déformables", Paris (1909).

[4] J.L. Ericksen and C. Truesdell: "Exact theory of stress and strain in rods and shells", Arch. Rat. Mech. Analysis 1, 295-323 (1958).

[5] J. Sudria: "L'action euclidienne de déformation et de mouvement", Mem. Sc. Physique, 29, Paris (1935).

[6] E. Kröner: "Kontinuumstheorie der Versetzungen und Eigenspannungen", Ergeb. Angew. Math., 5, Berlin-Göttingen-Heidelberg, (1958).

[7] W. Günther: "Zur Statik und Kinematik des Cosseratschen Kontinuums", Abh. Braunschw. Wiss. Ges. 10, 195-213 (1958).

[8] H. Schäfer: "Versuch einer Elastizitätstheorie der zweidimensionalen ebenen Cosserat-Kontinuums", Miszellaneen der angew. Mech., Festschrift W. Tolmien, 277-292, Berlin (1962).

[9] S.C. Cowin: "Mechanics of Cosserat continua", Doct. Diss. Pennsylvania State Univ., (1962).

[10] E. L. Aero and E. V. Kuvshinskii: "Osnovnie uravnenia teorii uprugosti sred s vrashchatal'nim vzaimodeistviem chastic", Fiz. Tv. Tela 2, 1399-1409, (1960).

[11] C. Truesdell and R. Toupin: "The classical field theories", Handb. der Phys. (ed. S. Flügge), Bd. III/1, Berlin-Götting.-Heidelberg, (1960).

[12] G. Grioli: "Elasticità asimmetrica", Ann. di Mat. Pura ed Appl., Ser. IV, 50, 389-417 (1960).

[13] R. Toupin: "Elastic materials with couple-stresses", Arch. Rat. Mech. Anal. 11, 385-414 (1962).

[14] E. V. Kuvshinskii, E. L. Aero: "Kontinual'naja teoria asimetricheskoi uprugosti. Uchet "vnutrennego" vrashchenia", Fiz. Tv. Tela 5, 2591-2598 (1963).

[15] R. Toupin: "Theories of elasticity with couple-stresses" Arch. Rat. Mech. Anal. 17, 85-112 (1964).

[16] A. C. Eringen: "Theory of micropolar continua", Proc. 9th Midwestern Mech. Conference, Madison, Wisconsin (1965).

[17] A. C. Eringen and E. S. Suhubi: "Nonlinear theory of simple microelastic solids", Int. J. Engng. Sc., 2, 189-203 (1964).

[18] R. Stojanovic and S. Djuric: "On the measures of strain in the theory of the elastic generalized Cosserat continua", Symposia Mathematics 1, 211-228, (1968).

[19] R. Stojanovic: "Mechanics of Polar Continua", CISM, Udine, (1969).

[20] C. B. Kafadar and A. C. Eringen: "Micropolar media-I. The classical theory", Int. J. Eng. Sc., 9, 271--305 (1971).

[21] R. Toupin: "The elastic dielectric", J. Rat. Mech. Anal. 5, 849-915 (1955).

W. NOWACKI

THE MICROPOLAR THERMOELASTICITY

1. Introduction

Thermoelasticity investigates the interaction of the field of deformation with the field of temperature and combines, on the basis of the thermodynamics of the irreversible processes, two separately developing branches of science, namely the theory of elasticity and the theory of heat conduction.

At the present moment, after 20 years of the development, the thermoelasticity of the Hooke's continuum is fully formed. The fundamental assumptions have been worked out [1] - [5], the fundamental relations and different equations have been elaborated, the fundamental energy and variational theorems obtained. The entire classical thermoelasticity has been formulated in a number of monographs.

On the background of the development of the classical thermoelasticity the achievements of Cosserat's continuum thermoelasticity [6 - 9], are still modest. Though all more important theorems have been derived, the domain of the particular solutions is incomparably smaller. The fundamentals of the micropolar, Cosserats' thermoelasticity were formulated in 1966 by the author of the present study [10], [11].

We present, in a concise form, the fundamental relations and the fundamental equations of Cosserats' continuum thermoelasticity. The principle of the energy conservation and the entropy balance are our point of departure

$$\frac{d}{dt}\int_V \left[\frac{1}{2}(\rho v_i v_i + I w_i w_i) + U\right] dV = \int_V (X_i v_i + Y_i w_i)\, dV +$$

(1.1) $$+ \int_A (p_i v_i + m_i w_i)\, dA - \int_A q_i n_i\, dA$$

and

(1.2) $$\int_V \dot{S}\, dV = -\int_A \frac{q_i n_i}{T}\, dA + \int_V \Theta\, dV .$$

In eq. (1.1) U denotes the internal energy referred to the unit of volume, X_i, Y_i are the components of the body forces and moments acting on the surface A bounding the body, u_i, φ_i denote the components of the displacement vector and rotation vector, respectively, $\dot{u}_i = v_i$, $\dot{\varphi}_i = w_i$ are their time derivatives, $\underline{q}$ is the flux of heat vector, ρ the density, I the rotational inertia.

The term on the left-hand side of eq. (1.1) represents the time change of the internal and kinetic energies. The first term on the right hand side of the equation is the power of body forces and body moments, the second term is the power of traction and surface moments. The last term expresses the amount of heat transmitted into the volume V by the heat conduction.

The left-hand side of the balance of entropy equation (1.2) represents the increase of entropy. The first term on the right-hand side is the increase of entropy due to the exchange of the entropy with environment, the second term expresses the production of entropy generated by heat conduction. Here $\Theta = -\frac{q_i T_{,i}}{T^2} > 0$, according to the postulate of the thermodynamics of irreversible processes.

In eq. (1.2) S denotes the entropy referred to the unit of volume, T is the absolute temperature, Θ is the source of entropy.

Transforming eqs. (1.1) and (1.2) by means of the equation of motion

$$\left.\begin{aligned} \sigma_{ji,j} + X_i &= \rho \ddot{u}_i , \\ \epsilon_{ijk}\sigma_{jk} + \mu_{ji,j} + Y_i &= I \ddot{\varphi}_i , \end{aligned}\right\} \tag{1.3}$$

where σ_{ji}, μ_{ji} are the components of force stress and moment stress tensors, ϵ_{ijk} is Ricci's alternator, and taking into account the definitions of the asymmetric strain tensors

$$\gamma_{ji} = u_{i,j} - \epsilon_{kji}\varphi_k \quad , \qquad \varkappa_{ji} = \varphi_{i,j} \tag{1.4}$$

we obtain, eliminating the quantity $q_{i,i}$ and introducing the free energy $F = U - ST$ the following equation

$$\dot{F} = \sigma_{ji}\dot{\gamma}_{ji} + \mu_{ji}\dot{\varkappa}_{ji} - \dot{T}S \,. \tag{1.5}$$

Since the free energy is the function of the independent variables γ_{ji}, $\varkappa_{ji}$ and T then

$$\dot{F} = \frac{\partial F}{\partial \gamma_{ji}}\dot{\gamma}_{ji} + \frac{\partial F}{\partial \varkappa_{ji}}\dot{\varkappa}_{ji} + \frac{\partial F}{\partial T}\dot{T} \,. \tag{1.6}$$

We assume that the functions Θ, q_i,.. σ_{ji}, μ_{ji} do not depend explicitly on the time derivatives of the functions γ_{ji}, $\varkappa_{ji}$, T next, defining $S = -\frac{\partial F}{\partial \gamma_{ji}}$ we obtain the following relations

$$\sigma_{ji} = \frac{\partial F}{\partial \gamma_{ji}} \,, \quad \mu_{ji} = \frac{\partial F}{\partial \varkappa_{ji}} \,, \quad S = -\frac{\partial F}{\partial T} \,. \tag{1.7}$$

Let us return to the inequality

$$\Theta = -\frac{T_{,i}\, q_i}{T^2} \geq 0 \,.$$

This inequality is satisfied by the Fourier's law of heat conduction

$$-q_i = k_{ij}\, T_{,j} \,. \tag{1.8}$$

Consequently, we obtain from the entropy balance, for a homogeneous and isotropic body,

$$T\dot{S} = -q_{i,i} = k\, T_{,jj} \,, \tag{1.9}$$

here k is the coefficient of the thermal conduction.

Let us expand the free energy $F(\gamma_{ji}, \varkappa_{ji}, T)$ into the Taylor series in the vicinity of the natural state ($\varkappa_{ji} = \gamma_{ji} = 0$, $T = T_0$), disregarding the terms of higher order than the second one. For isotropic, homogeneous, and centrosymmetric bodies, we obtain the following form of the expansion

$$F = \frac{\mu+\alpha}{2}\gamma_{ji}\gamma_{ji} + \frac{\mu-\alpha}{2}\gamma_{ji}\gamma_{ij} + \frac{\lambda}{2}\gamma_{kk}\gamma_{nn} + \frac{\gamma+\varepsilon}{2}\varkappa_{ji}\varkappa_{ji} + \frac{\gamma-\varepsilon}{2}\varkappa_{ji}\varkappa_{ij} + \frac{\beta}{2}\varkappa_{kk}\varkappa_{nn} - \nu\gamma_{kk}\vartheta - \frac{m}{2}\vartheta^2 \tag{1.10}$$

Here $\vartheta = T - T_0$ where T_0 is the temperature of the natural state, the magnitudes $\mu, \alpha, \lambda, \gamma, \varepsilon, \beta$ denote the material constants ν, m are the quantities containing the mechanical and thermal constants. On the right-hand side there occurs the independent quadratic invariants $\gamma_{ji}\gamma_{ji}$, $\gamma_{ji}\gamma_{ij}$, $\gamma_{kk}\gamma_{nn}$ and the invariant γ_{kk}. The quantities $\gamma_{ji}\varkappa_{ji}$, $\gamma_{ji}\varkappa_{ij}$, $\gamma_{kk}\varkappa_{nn}$, $\varkappa_{nn}\vartheta$ cannot enter the equation because of the assumption of the centrosymmetry.

Consequently, making use of eqs. (1.6), we obtain the following constitutive equations

$$\left.\begin{aligned} \sigma_{ji} &= (\mu+\alpha)\gamma_{ji} + (\mu-\alpha)\gamma_{ij} + (\lambda\gamma_{kk} - \nu\vartheta)\delta_{ij}, \\ \mu_{ji} &= (\gamma+\varepsilon)\varkappa_{ji} + (\gamma-\varepsilon)\varkappa_{ij} + \beta\varkappa_{kk}\delta_{ij}, \\ S &= \nu\gamma_{kk} + m\vartheta = \nu\gamma_{kk} + \frac{c_\varepsilon}{T_0}\vartheta . \end{aligned}\right\} \tag{1.11}$$

Here $\mu, \lambda, \alpha, \beta, \gamma, \varepsilon$ are the material constants in the isothermal state, $\nu = (3\lambda + 2\mu)\alpha_t$ where α_t is the coefficient of the linear thermal expansion, while c_ε is the specific heat at constant deformation. We should remark that the constitutive equation $(1.11)_3$ holds true only for the limitation $\left|\frac{\vartheta}{T_0}\right| \ll 1$. From the entropy balance

$$TS\dot{} = -q_{i,i} + W \tag{1.12}$$

where W denotes the amount of heat generated in a unit of volume and time, from the Fourier law $q_i = -kT_{,i}$ and from the equation $(1.11)_3$, we are lead to the lineat heat conduction equation

$$\nabla^2\vartheta - \frac{1}{\varkappa}\dot{\vartheta} - \eta \operatorname{div}\underline{\dot{u}} = -\frac{Q}{\varkappa}\,, \quad Q = \frac{W}{c_\varepsilon}\,, \quad \vartheta = T - T_0\,. \tag{1.13}$$

2. The Dynamical Problem of Thermoelasticity

Let us consider a regular region of space $V+A$ where A is the boundary containing an elastic, homogeneous, isotropic and centrosymmetric micropolar continuum.

Let $\sigma_{ji}(\underline{x},t)$ be the components of the non-symmetric force stress tensor, while $\mu_{ji}(\underline{x},t)$ the components of the non-symmetric moment stress tensor $\underline{u}(\underline{x},t)$ denotes the displacement vector and $\underline{\varphi}(\underline{x},t)$ is the vector of rotation. By $\vartheta = T - T_0$ we denote the change of temperature.

The dynamical problem of thermoelasticity consists in determining the functions

$$\sigma_{ji}(\underline{x},t)\,,\ \mu_{ji}(\underline{x},t)\,,\ \gamma_{ji}(\underline{x},t)\,,\ \varkappa_{ji}(\underline{x},t)\,,\ \underline{u}(\underline{x},t)\,,\ \underline{\varphi}(\underline{x},t)$$

$$\text{and}\quad \vartheta(\underline{x},t)\ \text{for}\ \underline{x}\in V+A\,.$$

These functions ought to satisfy:

a) the equation of motion,

b) the equation of thermal conduction,

c) the linearized constitutive equations,

d) the boundary conditions

$$\begin{cases} \mu_{ji} n_j = m_i(\underline{x}, t), \; \sigma_{ji} n_j = p_i(\underline{x}, t), \; \theta = \vartheta(\underline{x}, t), \; \underline{x} \in A_\sigma, \; t > 0, \\ u_i = f_i(\underline{x}, t), \; \varphi_i = g_i(\underline{x}, t), \; \theta = \vartheta(\underline{x}, t), \; \underline{x} \in A_u, \; t > 0. \end{cases} \tag{2.1}$$

e) the initial conditions

$$\left.\begin{aligned} u_i = k_i(\underline{x}), \quad \dot{u}_i = h_i(\underline{x}), \quad \theta = s(\underline{x}), \\ \varphi_i = l_i(\underline{x}), \quad \dot{\varphi}_i = s_i(\underline{x}), \quad \underline{x} \in V, \; t = 0. \end{aligned}\right\} \tag{2.2}$$

The functions $p_i, m_i, f_i, g_i, \vartheta$ in the boundary conditions and k_i, h_i s, l_i, s_i in the initial conditions are given functions.

Let us pass to the representation of the differential equations of the problem choosing as unknown functions the displacements $\underline{u}(\underline{x},t)$, the rotations $\underline{\varphi}(\underline{x},t)$ and the temperature $\theta(\underline{x},t)$. Eliminating the stresses from the equations of motion by means of constitutive equations, expressing them by the functions $\underline{u}$ and $\underline{\varphi}$, we obtain, together with the equation of heat conduction, the following set of differential equations of thermoelasticity.

$$(2.3)\qquad \begin{cases} \Box_2 \underline{u} + (\lambda + \mu - \alpha)\,\mathrm{grad\,div}\,\underline{u} + 2\alpha\,\mathrm{curl}\,\underline{\varphi} + \underline{X} = \nu\,\mathrm{grad}\,\vartheta\,, \\ \Box_4 \underline{\varphi} + (\beta + \gamma - \varepsilon)\,\mathrm{grad\,div}\,\underline{\varphi} + 2\alpha\,\mathrm{curl}\,\underline{u} + \underline{Y} = 0\,, \\ \nabla^2\vartheta - \frac{1}{\varkappa}\dot{\vartheta} - \eta\,\mathrm{div}\,\dot{\underline{u}} = -\frac{Q}{\varkappa}\,, \end{cases}$$

where $\Box_2 = (\mu + \alpha)\nabla^2 - \rho\partial_t^2$, $\Box_4 = (\varepsilon + \gamma)\nabla^2 - 4\alpha - J\partial_t^2$.

We have obtained a coupled system of seven equations for seven unknowns, namely three components of the displacement $\underline{u}$ three components of the rotation $\underline{\varphi}$ and the temperature ϑ . These fields can be generated by loadings, a surface heating, body forces and moments and heat sources.

In eqs.(2.3) the mutually independent functions $\underline{u}, \underline{\varphi}, \vartheta$ are coupled. The change of deformation field generates the change generates the change of temperature and vice versa.

The coupled system of equations (2.3) is complicated and inconvenient to deal with; hence, our prime objective will be to uncouple it.

The dynamic equations of thermoelasticity

$$(2.4)\qquad \begin{cases} \Box_2 \underline{u} + (\lambda + \mu - \alpha)\,\mathrm{grad\,div}\,\underline{u} + 2\alpha\,\mathrm{curl}\,\underline{\varphi} + \underline{X} = \nu\,\mathrm{grad}\,\vartheta\,, \\ \Box_4 \underline{\varphi} + (\beta + \gamma - \varepsilon)\,\mathrm{grad\,div}\,\underline{\varphi} + 2\alpha\,\mathrm{curl}\,\underline{u} + Y = 0 \\ D\vartheta - \eta\,\mathrm{div}\,\dot{\underline{u}} = -\frac{Q}{\varkappa}\,, \quad D = \nabla^2 - \frac{1}{\varkappa}\partial_t\,, \end{cases}$$

can be separated in two different ways. The first way, analogous to Lame's procedure applied in classical elastokinetics, consists in the decomposition of the vectors $\underline{u}$ and $\underline{\varphi}$ into potential and solenoidal parts, respectively

$$\left.\begin{aligned} \underline{u} &= \operatorname{grad}\Phi + \operatorname{curl}\underline{\Psi} , & \operatorname{div}\underline{\Psi} &= 0 , \\ \underline{\varphi} &= \operatorname{grad}\Gamma + \operatorname{curl}\underline{H} , & \operatorname{div}\underline{H} &= 0 . \end{aligned}\right\} \qquad (2.5)$$

Decomposing in a similar way the expressions for the body forces and the body couples

$$\left.\begin{aligned} \underline{X} &= \rho(\operatorname{grad}\vartheta + \operatorname{curl}\underline{\chi}) , & \operatorname{div}\underline{\chi} &= 0 , \\ \underline{Y} &= J(\operatorname{grad}\sigma + \operatorname{curl}\underline{\eta}) , & \operatorname{div}\underline{\eta} &= 0 , \end{aligned}\right\} \qquad (2.6)$$

and substituting (2.5) and (2.6) to the equations of thermoelasticity (2.4), we obtain the following system of equations

$$\left.\begin{aligned} &\Box_1\Phi + \rho\vartheta = \nu\theta , \\ &D\theta - \eta\partial_t\nabla^2\dot{\Phi} = -\frac{Q}{\varkappa} , \\ &\Box_3\Gamma + J\sigma = 0 , \\ &\Box_2\underline{\Psi} + 2\alpha\operatorname{curl}\underline{H} + \rho\underline{\chi} = 0 , \\ &\Box_4\underline{H} + 2\alpha\operatorname{curl}\underline{\Psi} + J\underline{\eta} = 0 , \\ &D = \nabla^2 - \frac{1}{\varkappa}\partial_t , \\ &\Box_3 = \nabla^2 - \frac{1}{c_3^2}\partial_t^2 - 4\alpha . \end{aligned}\right\} \qquad (2.7)$$

The complex system of equations of thermoelasticity has been reduced to the solution of simple wave equations (2.7). Eq. $(2.7)_1$ represents a longitudinal wave, $(2.7)_2$ the heat conduction equation, eq. $(2.7)_3$ the longitudinal micro-rotational wave, eq. $(2.7)_4$, $(2.7)_5$ correspond to a transversal displacement and transversal micro-rotational wave.

The form of eqs. $(2.7)_1$ and $(2.7)_2$ is identical with the form of the longitudinal wave equation of the classical thermoelasticity; on the contrary, eq. $(2.7)_3$ is a new one, namely the Klein-Gordon differential equation. Let us notice that eqs. $(2.7)_1$, $(2.7)_2$ and eqs. $(2.7)_4$, $(2.7)_5$ are mutually coupled.

After elimination, we obtain

$$
(2.8)\quad
\begin{cases}
(\Box_1 D - \nu\eta\partial_t\nabla^2)\Phi = -\dfrac{\nu}{\varkappa}Q - \rho D\vartheta \,, \\[2mm]
(\Box_1 D - \nu\eta\partial_t\nabla^2)\theta = -\dfrac{\Box_1^2 Q}{\varkappa} - \rho\eta\partial_t\nabla^2\vartheta \,, \\[2mm]
\Box_3\Gamma + J\sigma = 0 \,, \\[2mm]
(\Box_2\Box_4 + 4\alpha^2\nabla^2)\underline{\Psi} = 2\alpha\,\operatorname{curl}\underline{\eta} - \rho\,\Box_4\underline{\chi} \,, \\[2mm]
(\Box_2\Box_4 + 4\alpha^2\nabla^2)\underline{H} = 2\alpha\rho\operatorname{curl}\underline{\chi} - J\,\Box_2\underline{\eta} \,.
\end{cases}
$$

We shall consider first the propagation of thermoelastic waves in an unbounded space.

If the quantities $\sigma, \underline{\chi}, \underline{\eta}$ and the initial conditions of the functions $\Gamma, \underline{\Psi}, \underline{H}$ are equal to zero, then in an unbounded elastic space only dilatational waves will propagate. Eq. $(2.8)_1$ describing the waves is identical with that obtained for the elas tic classical medium (with no couple-stresses). These waves are attenuated and dispersed.

Since

$$u_i = \Phi_{,i}\,,\quad \varphi_i = 0\,,\quad \gamma_{ji} = \Phi_{,ji}\,,\quad \varkappa_{ji} = 0\,, \tag{2.9}$$

we have

$$\sigma_{ij} = 2\mu(\Phi_{,ij} - \delta_{ij}\Phi_{,kk}) + \rho\delta_{ij}(\ddot{\Phi} - \vartheta)\,, \qquad \sigma_{\langle ij\rangle} = 0. \tag{2.10}$$

If $Q, \vartheta, \underline{\chi}, \underline{\eta}$ are equal to zero and the initial conditions of the functions $Q, \Phi, \underline{\Psi}, \underline{H}$ are homogeneous, then in an infinite medium only microrotational waves, described by eq. $(2.8)_3$ propagate. We have namely

$$u_i = 0\,,\quad \varphi_i = \Gamma_{,i}\,,\quad \varkappa_{ji} = \Gamma_{,ij}\,,\quad \gamma_{(ji)} = 0\,,\quad \gamma_{\langle ji\rangle} = -\epsilon_{kij}\Gamma_{,k}\,. \tag{2.11}$$

Couple-stresses and force-stresses will appear in the medium forming the symmetric tensor μ_{ij}

$$\mu_{(ij)} = 2\gamma\Gamma_{,ij} + \beta\delta_{ij}\Gamma_{,kk}\,,\quad \mu_{\langle ij\rangle} = 0\,,\quad \sigma_{ij} = 0\,,\quad \operatorname{div}\underline{u} = 0\,, \tag{2.12}$$

and the asymmetric tensor σ_{ij}

$$\sigma_{\langle ij\rangle} = 2\mu\,\gamma_{\langle ij\rangle}\,,\qquad \sigma_{(ij)} = 0\,.$$

The propagation of these waves is not accompanied by a temperature field.

Finally, in the case when the quantities Q, ϑ, σ are equal to zero and the initial conditions of the functions

Φ, Γ, Q are homogeneous, only transverse waves propagate in an infinite space (described by the eqs. $(2.8)_4$ and $(2.8)_5$). In an infinite medium these waves are not accompanied by any temperature field. Since $\operatorname{div} \underline{u} = 0$ they do not induce any changes in the volume of the body.

In a finite medium all three kinds of waves discussed here appear. Eqs. $(2.8)_1$ and $(2.8)_5$ are coupled by means of the boundary conditions.

The second method of separation of the system (2.10) is analogous to that applied by Galerkin [12] to the classical elastostatics and by M. Iacovache to the classical elastokinetics [13]. Functions of this type, suitable for asymmetric elasticity were established by N. Sandru [14].

The functions of this kind for the case of the dynamic problems of thermoelasticity have been devised by W. Nowacki [15]. These functions have been derived by another method by J. Stefaniak [16].

Below we give the final result of the separation. We represent the vectors $\underline{u}, \underline{\varphi}$ and the temperature ϑ by means of two vector functions $\underline{F}$, $\underline{G}$ and one scalar function L

$$(2.13)\quad \begin{cases} \underline{u} = \Box_4 M\underline{F} - \operatorname{grad}\operatorname{div} N\underline{F} - 2\alpha \operatorname{curl} \Box_3 \underline{G} + \gamma \operatorname{grad} L\,, \\ \underline{\varphi} = \Box_2 \Box_3 \underline{G} - \operatorname{grad}\operatorname{div} \Theta \underline{G} - 2\alpha \operatorname{curl} M\underline{F}\,, \\ \vartheta = \eta \operatorname{div} \partial_t \Omega \underline{F} + \Box_1 L\,, \end{cases}$$

where

$$\Omega = \Box_2\Box_4 + 4\alpha^2\nabla^2 \quad , \qquad M = \Box_1 D - \nu\eta\partial_t\nabla^2 ,$$

$$N = \Gamma D - \nu\eta\partial_t\Box_4 \quad , \qquad \Gamma = (\lambda + \mu - \alpha)\Box_4 - 4\alpha^2 ,$$

$$\Theta = (\beta + \gamma - \varepsilon)\Box_2 - 4\alpha^2 .$$

Inserting the relations (2.13) into the system of eqs. (2.4) we obtain the following wave repeated equations

$$\left.\begin{aligned} &\Omega M \underline{F} + \underline{X} = 0, \\ &\Omega \Box_3 \underline{G} + \underline{Y} = 0, \\ &M L + Q/\varkappa = 0. \end{aligned}\right\} \qquad (2.14)$$

We have obtained a system of equations in which the body forces, the body couples and heat sources appear separately. Let us notice, that in an infinite, elastic space, the assumption $\underline{X} = 0$ with homogeneous initial conditions for $\underline{F}$, leads to the conclusion that $\underline{F} = 0$ in the whole space. The same result holds for the function $\underline{G} = 0$ with $\underline{Y} = 0$ and $L = 0$ with $Q = 0$.

Eqs. (2.14) are particularly useful in the case of the singular solutions for an infinite, micropolar space. Such solutions have been obtained for the case of the action of concentrated forces, concentrated moments and heat sources harmonically varying in time [17] . Then eqs. (2.14) simplify to a simple system of elliptic equations.

In this case we obtain the system of equations

$$
(2.15)\quad \begin{cases} A(\nabla^2+k_1^2)(\nabla^2+k_2^2)(\nabla^2+\mu_1^2)(\nabla^2+\mu_2^2)\underline{F}^*+\underline{X}^*=0\,, \\ B(\nabla^2+k_1^2)(\nabla^2+k_2^2)(\nabla^2+k_3^2)\underline{G}^*+\underline{Y}^*=0\,, \\ C(\nabla^2+m^2)(\nabla^2+\mu_2^2)L^*+Q^*/\varkappa=0\,, \end{cases}
$$

where

$$\underline{X}=\underline{X}^*(\underline{x})e^{-i\omega t}\,,\qquad \underline{Y}=\underline{Y}^*(\underline{x})e^{-i\omega t}\,,\quad \text{and so on.}$$

Let us return to the fundamental equations of thermoelasticity (2.4). Passing to the cylindrical coordinate system (r, ϑ, z) and assuming that the deformation possesses the axial symmetry, we obtain two independent systems of equations.

In the first system of equations the following components of the vectors $\underline{u}$, $\underline{\varphi}$, $\underline{X}$, $\underline{Y}$ grad θ, appear

$$
(2.16)\quad \begin{gathered} \underline{u}=(u_r,0,u_z)\,,\quad \underline{\varphi}=(0,\varphi_\vartheta,0)\,,\quad \underline{X}=(X_r,0,X_z)\,,\quad \underline{Y}=(0,Y_\vartheta,0)\,, \\ \operatorname{grad}\theta=\left(\frac{\partial\theta}{\partial r},\,0,\,\frac{\partial\theta}{\partial z}\right). \end{gathered}
$$

Now the system of equations has the form

$$
(2.17)\quad \begin{cases} (\mu+\alpha)\left(\nabla^2-\frac{1}{r^2}\right)u_r+(\lambda+\mu-\alpha)\frac{\partial e}{\partial r}-2\alpha\frac{\partial\varphi_\vartheta}{\partial z}+X_r=\nu\frac{\partial\theta}{\partial r}+\varrho\ddot{u}_r\,, \\ (\mu+\alpha)\nabla^2u_z+(\lambda+\mu-\alpha)\frac{\partial e}{\partial z}+2\alpha\frac{1}{r}\frac{\partial}{\partial r}(r\varphi_\vartheta)+X_z=\nu\frac{\partial\theta}{\partial z}+\varrho\ddot{u}_z\,, \\ (\gamma+\varepsilon)\left(\nabla^2-\frac{1}{r^2}\right)\varphi_\vartheta-4\alpha\varphi_\vartheta+2\alpha\left(\frac{\partial u_r}{\partial z}-\frac{\partial u_z}{\partial r}\right)+Y_\vartheta=J\ddot{\varphi}_\vartheta\,, \end{cases}
$$

$$\left(\nabla^2 - \frac{1}{r^2}\right)\vartheta - \frac{1}{\varkappa}\dot{\vartheta} - \eta\partial_t e = -\frac{Q}{\varkappa} \tag{2.17}$$

where

$$e = \frac{1}{r}\frac{\partial}{\partial r}(u_r r) + \frac{\partial u_z}{\partial z}\ , \qquad \nabla^2 = \frac{\partial^2}{\partial r^2} + \frac{1}{r}\frac{\partial}{\partial r} + \frac{\partial^2}{\partial z^2}\ .$$

In the second system of equations, characterized by the vectors

$$\underline{u} \equiv (0, u_\vartheta, 0)\ ,\ \underline{\varphi} \equiv (\varphi_r, 0, \varphi_z)\ ;\ \underline{X} \equiv (0, X_\vartheta, 0)\ ,\ \underline{Y} \equiv (Y_r, 0, Y_z)\ , \tag{2.18}$$

the thermal terms vanish, therefore the vectors $\underline{u}, \underline{\varphi}$ do not depend on the field of temperature.

The system of equations takes the following form

$$\left.\begin{aligned}
&(\gamma+\varepsilon)\left(\nabla^2 - \frac{1}{r^2}\right)\varphi_r - 4\alpha\varphi_r + (\beta+\gamma-\varepsilon)\frac{\partial \varkappa}{\partial r} - 2\alpha\frac{\partial u_\vartheta}{\partial z} + Y_r = I\ddot{\varphi}_r\,,\\
&(\gamma+\varepsilon)\nabla^2\varphi_z - 4\alpha\varphi_z + (\beta+\gamma-\varepsilon)\frac{\partial \varkappa}{\partial z} + 2\frac{\alpha}{r}\frac{\partial}{\partial r}(r u_\vartheta) + Y_z = J\ddot{\varphi}\ ,\\
&(\mu+\alpha)\left(\nabla^2 - \frac{1}{r^2}\right)u_\vartheta + 2\alpha\left(\frac{\partial \varphi_r}{\partial z} - \frac{\partial \varphi_z}{\partial r}\right) + X_\vartheta = \rho\ddot{u}_\vartheta\ .
\end{aligned}\right\} \tag{2.19}$$

Similarly in the two-dimensional state of stress the system of equations (2.4) can be split into two independent sets of equations. Under the assumption that the deformation of a body does not depend on the independent variable x_3 the following vectors

$$\underline{u} = (u_1, u_2, 0)\ ,\ \underline{\varphi} = (0, 0, \varphi_3)\ ,\ \underline{X} = (X_1, X_2, 0)\ ,$$
$$\underline{Y} = (0, 0, Y_3)\ ,\quad \operatorname{grad}\vartheta = (\partial_1\vartheta, \partial_2\vartheta, 0)\ , \tag{2.20}$$

occur in the first system.

Here the system of equations has the form

$$\begin{cases} [(\mu+\alpha)\nabla_1^2 - \varrho\partial_t^2]u_1 + (\lambda+\mu-\alpha)\partial_1 e + 2\alpha\partial_2\varphi_3 + X_1 = \nu\partial_1\vartheta\ , \\ [(\mu+\alpha)\nabla_1^2 - \varrho\partial_t^2]u_2 + (\lambda+\mu-\alpha)\partial_2 e - 2\alpha\partial_1\varphi_3 + X_2 = \nu\partial_1\vartheta\ , \\ [(\gamma+\varepsilon)\nabla_1^2 - 4\alpha - J\partial_t^2]\varphi_3 + 2\alpha(\partial_1 u_2 - \partial_2 u_1) + Y_3 = 0\ . \end{cases} \tag{2.21}$$

$$\left(\nabla^2 - \frac{1}{\varkappa}\partial_t\right)\vartheta - \eta\partial_t e = -Q/\varkappa\ ,\quad e = \partial_1 u_1 + \partial_2 u_2\ .$$

The second system in which the components of the vectors

$$\underline{u} = (0, 0, u_3)\ ,\ \underline{\varphi} = (\varphi_1, \varphi_2, 0)\ ,\ \underline{X} = (0, 0, X_3)\ ,\ \underline{Y} = (Y_1, Y_2, 0)\ , \tag{2.22}$$

appear, is independent of the field of temperature.

We have already noticed that for an infinite space and P-wave the results for the micropolar and Hooke's continua are of the same form.

Thus the following question arises: does the same situation occur for certain bounded bodies? It is easy to observe that such cases concern the one-dimensional problems.

Thus if all the causes and effects depend only on the variables x_1 and t, then we obtain the following system of equations

$$\left.\begin{aligned} &[(\lambda+2\mu)\partial_1^2-\varrho\partial_t^2]u_1=\nu\partial_1\theta\,,\\ &\left(\partial_1^2-\frac{1}{\varkappa}\partial_t\right)\theta-\eta\partial_t\partial_1 u_1=-\frac{Q}{\varkappa}\,,\end{aligned}\right\}\qquad(2.23)$$

which is exactly of the same form for Hooke's and micropolar media.

If all the causes depend on the variables $r=(x_1^2+x_2^2)^{1/2}$ and t then the system of coupled equations

$$\left.\begin{aligned} &(\lambda+2\mu)\left(\nabla_r^2-\frac{1}{r^2}\right)u_r-\varrho\ddot{u}_r=\nu\frac{\partial\theta}{\partial r}\,,\\ &\left(\nabla_1^2-\frac{1}{\varkappa}\partial_t\right)\theta-\eta\partial_t\left(\frac{1}{r}\frac{\partial}{\partial r}(r u_r)\right)=-\frac{Q}{\varkappa}\,,\\ &\nabla_r^2=\frac{\partial^2}{\partial r^2}+\frac{1}{r}\frac{\partial}{\partial r}\,,\end{aligned}\right\}\qquad(2.24)$$

has the same form both for Hooke's and micropolar media. Likewise the equations

$$\begin{aligned} &(\lambda+2\mu)\left(\nabla_R^2-\frac{1}{R^2}\right)u_R-\varrho\ddot{u}_R=\nu\frac{\partial\theta}{\partial R}\,,\\ &\left(\nabla_R^2-\frac{1}{\varkappa}\partial_t\right)\theta-\eta\partial_t\frac{1}{R^2}\frac{\partial}{\partial R}(u_R R^2)=-\frac{Q}{\varkappa}\,,\quad \nabla_R^2=\frac{\partial^2}{\partial R^2}+\frac{2}{R}\frac{\partial}{\partial r}\,,\end{aligned}\qquad(2.25)$$

determining the strain symmetric with respect to a point, have the same form for both the media.

So far all more important general theorems have been derived, let us mention here the principle of virtual work, the theorems of energy, the reciprocity theorem, the generalized formulae of Somigliano and Green. We shall not dwell on these problems referring to the bibliography [10].

We have given a short review of the results of the dynamical thermoelasticity. The theory can be considerably simplified by neglecting the coupling of the thermal conduction equation with the equations in displacements and rotations. We disregard the term $-\eta \operatorname{div} \dot{u}$ in the heat conduction equation. This simplification does not influence the magnitude of stresses and strains, however it does change the qualitative picture of the problem. From the wave equation $(2.8)_1$ it results that the P-wave is attenuated and dispersed; on the other hand when we disregard the term $\eta \operatorname{div} \dot{u}$ the P-wave consists only of the elastic part (which is neither attenuated nor dispersed) and the diffusive wave. Besides, we cannot obtain any information on the amount of the dissipated energy from the simplified theory.

In recent years the dynamical problems of thermoelasticity have been extended on the Cosserats' continuum of viscoelastic properties. D. Iesan [18] has given a few general theorems for such a continuum (the reciprocity theorem, the variational theorems and the uniqueness theorem). He also examined with full particulars the plane dynamical problem [25]. S. Kaliski [19] has given the fundamentals of the thermo-magneto-microelasticity. By this term we understand the coupling of the field of deformation (of the Cosserats' continuum) with the field of temperature and the electromagnetic field in the conductors.

3. Stationary Thermal Stress Problems

For a stationary heat flow the time derivative in the thermoelasticity equations disappears and all the quantities take the following form

$$\left.\begin{aligned}&(\mu+\alpha)\nabla^2\underline{u}+(\lambda+\mu-\alpha)\,\mathrm{grad}\,\mathrm{div}\,\underline{u}+2\alpha\,\mathrm{curl}\,\underline{\varphi}=\nu\,\mathrm{grad}\,\vartheta,\\&(\gamma+\varepsilon)\nabla^2\underline{\varphi}-4\alpha\underline{\varphi}+(\beta+\gamma-\varepsilon)\,\mathrm{grad}\,\mathrm{div}\,\underline{\varphi}+2\alpha\,\mathrm{curl}\,\underline{u}=0,\\&\nabla^2\vartheta=-\frac{Q}{\varkappa}\;.\end{aligned}\right\}\qquad(3.1)$$

The equation of thermal conduction has become an equation of the Poisson type. We determine the temperature Q from eq. $(3.1)_3$ and substitute it, as a known function, in eq. $(3.1)_1$. Only eqs. $(3.1)_1$ and $(3.1)_2$ are coupled. The easiest way to obtain the solution to the system of equations is to introduce the potential of thermoelastic displacement

$$\underline{u}'=\mathrm{grad}\,\Phi \qquad (3.2)$$

and to assume that $\underline{\varphi}'=0$.

Substituting (3.2) in eq. $(3.1)_1$ and $(3.1)_2$ we reduce the system of equations to the Poisson equation

$$(3.3) \qquad \nabla^2\Phi = m\theta\,, \quad m = \frac{\nu}{\lambda+2\mu}\,, \quad \underline{\varphi}' = 0$$

We express the stresses by means of the function Φ :

$$(3.4) \qquad \sigma'_{ij} = \sigma'_{ji} = 2\mu(\Phi_{,ij} - \delta_{ij}\nabla^2\Phi) \qquad , \qquad \mu'_{ij} = 0$$

These quantities constitute the complete solution for an infinite space and are identical for Hooke's medium and for Cosserats' medium. If the region is bounded we add to the stresses σ'_{ij} , μ'_{ij} such stresses σ''_{ij} , μ''_{ij} that all the boundary conditions are satisfied. The stresses σ''_{ij} , μ''_{ij} are connected with such state of displacement and rotations $\underline{u}''$, $\underline{\varphi}''$ that satisfies the following homogeneous system of equations

$$(3.5) \qquad \begin{cases} (\mu+\alpha)\nabla^2\underline{u}'' + (\lambda+\mu-\alpha)\,\mathrm{grad\,div}\,\underline{u}'' + 2\alpha\,\mathrm{curl}\,\underline{\varphi}'' = 0\,, \\ (\gamma+\varepsilon)\nabla^2\underline{\varphi}'' - 4\alpha\underline{\varphi}'' + (\beta+\gamma-\varepsilon)\,\mathrm{grad\,div}\,\underline{\varphi}'' + 2\alpha\,\mathrm{curl}\,\underline{u}'' = 0. \end{cases}$$

Similarly as in the classical thermoelasticity here too we have the "body force analogy" [20], [21]. The principle of the virtual work is our point of departure

$$(3.6) \qquad \int_V (X_i\delta u_i + Y_i\delta\varphi_i)\,dV + \int_A (p_i\delta u_i + m_i\delta\varphi_i)\,dA = \int_V (\sigma_{ji}\delta\gamma_{ji} + \mu_{ji}\delta\varkappa_{ji})\,dV$$

It says that the virtual work of the external forces (body forces and moments, tractions and surface moments) is equal to the virtual work of internal forces. Substituting the generalized Duhamel-Neumann equations to the right-hand side of the equations we obtain

$$\int_V (X_i \delta u_i + Y_i \delta \varphi_i) dV + \int_A (p_i \delta u_i + m_i \delta \varphi_i) dA = \\ = \delta \mathcal{H} - \nu \int_V \Theta \delta \gamma_{kk} dV \tag{3.7}$$

where

$$\mathcal{H} = \int W dV \,, \quad W = \mu \gamma_{(ij)} \gamma_{(ij)} + \frac{\lambda}{2} \gamma_{kk} \gamma_{nn} + \alpha \gamma_{<ij>} \gamma_{<ij>} + \\ + \gamma \varkappa_{(ij)} \varkappa_{(ij)} + \frac{\beta}{2} \varkappa_{kk} \varkappa_{nn} + \varepsilon \varkappa_{<ij>} \varkappa_{<ij>} \,.$$

Eq. (3.7) can be represented in the form

$$\delta \mathcal{H} = \int [(X_i - \nu \Theta_{,i}) \delta u_i + Y_i \delta \varphi_i] dV + \\ + \int_A [(p_i + \nu \Theta n_i) \delta u_i + m_i \delta \varphi_i] dA \,. \tag{3.8}$$

Now we shall consider an identical body (i.e., of the same form and material), but placed under isothermal conditions. Let the body-forces X_i^* and the body couples Y_i^* act on the body. The tensions p_i^* and moments m_i^* are assumed to be given on the surface A_σ, while displacement u_i^* and rotations φ_i^* on A_u. We ask the following question: what should be the quantities X_i^* and Y_i^* - expressing forces and couples acting inside the body - and, on the other hand, the quantities p_i^* and m_i^* - expressing the tensions and moments acting on the surface A_σ - with identical boundary conditions for A_u in order to obtain the same field of displacement u_i and rotations φ_i in both. viz., thermoelastic and isothermal problems. To get the answer, we shall compare (3.8) with the virtual work equation

$$\delta \mathcal{K}_t = \int_V (X_i^* \delta u_i + Y_i^* \delta \varphi_i) dV + \int_A (p_i^* \delta u_i + m_i^* \delta \varphi_i) dA . \tag{3.9}$$

In view of the identity of u_i and φ_i fields, the left-hand parts of eqs. (3.8) and (3.9) are identical too; thus, we obtain the following relations

$$\begin{cases} X_i^* = X_i - \nu\Theta_{,i} , \quad Y_i^* = Y_i , \quad \underline{x} \in V , \\ p_i^* = p_i + \nu\Theta n_i , \quad m_i^* = m_i , \quad \underline{x} \in A_\sigma , \\ u_i^* = u_i , \quad \varphi_i^* = \varphi_i , \quad \underline{x} \in A_u . \end{cases} \tag{3.10}$$

Relations (3.10) represent the body forces analogy by means of which each steady-state problem can be reduced to the isothermal problem of the theory of asymmetric elasticity.

Now we may ask the question whether the solution to the stationary equations of thermoelasticity can be combined from two parts, the first of which is identical in the form with the solution of the classical thermoelasticity. The answer to this question is affirmative.

Following H. Schaefer [22], let us introduce

the vector

$$\underline{\zeta} = \frac{1}{2} \operatorname{curl} \underline{u} - \underline{\varphi} \tag{3.11}$$

and eliminate the function $\underline{\varphi}$ from the system of thermoelasticity equations. We obtain

$$\left.\begin{aligned} &\mu \nabla^2 \underline{u} + (\lambda + \mu) \operatorname{grad} \operatorname{div} \underline{u} - \nu \operatorname{grad} \theta = 2\alpha \operatorname{curl} \underline{\zeta} \ , \\ &[(\gamma + \varepsilon)\nabla^2 - 4\alpha]\underline{\zeta} + (\beta + \gamma - \varepsilon) \operatorname{grad} \operatorname{div} \underline{\zeta} = \frac{1}{2}(\gamma + \varepsilon)\nabla^2 \operatorname{curl} \underline{u} . \end{aligned}\right\} \tag{3.12}$$

Let us combine the solution of this system of equations from two parts

$$\underline{u} = \underline{u}' + \underline{u}'' \ , \quad \underline{\zeta} = \underline{\zeta}' + \underline{\zeta}'' \ , \quad \underline{\zeta}' = 0 \ .$$

Thus the system of equations (3.12) is split into two systems of equations

$$\mu \nabla^2 \underline{u}' + (\lambda + \mu) \operatorname{grad} \operatorname{div} \underline{u}' = \nu \operatorname{grad} \theta , \ \nabla^2 \operatorname{curl} \underline{u}' = 0 \ , \tag{3.13}$$

and

$$\left.\begin{aligned} &\mu \nabla^2 \underline{u}'' + (\lambda + \mu) \operatorname{grad} \operatorname{div} \underline{u}'' = 2\alpha \operatorname{curl} \underline{\zeta}'' \ , \\ &[(\gamma + \varepsilon)\nabla^2 - 4\alpha]\underline{\zeta}'' + (\beta + \gamma - \varepsilon) \operatorname{grad} \operatorname{div} \underline{\zeta}'' = \frac{1}{2}(\gamma + \varepsilon)\nabla^2 \operatorname{curl} \underline{u} . \end{aligned}\right\} \tag{3.14}$$

Let us note that the system of equations (3.13) is identical with the corresponding system of classical thermoelasticity. (However, the constants μ, λ occurring in eqs.

(3.13) refer to the micropolar body).

Therefore the solution $\mathbf{v}'$ can be taken from the classical thermoelasticity. This solution satisfies eq.(3.13) and the corresponding boundary conditions. Since $\underline{\zeta}' = 0$, which is equivalent to the assumption that $\gamma'_{\langle ij\rangle} = 0$, the tensor γ_{ij} is symmetric. Also the force stress tensor is symmetric

$$(3.15) \qquad \sigma'_{ij} = 2\mu\gamma_{(ij)} + (\lambda\gamma'_{kk} - \nu\theta)\delta_{ij} .$$

However the assumption $\underline{\zeta}' = 0$ is equivalent to the assumption $\underline{\varphi}' = \frac{1}{2}\,\mathrm{curl}\,\underline{u}'$. Since $\underline{\varphi}' \neq 0$ then there exists the tensor $\varkappa'_{ji} = \varphi'_{i,j}$ and the moment stresses

$$(3.16) \qquad \mu'_{ji} = 2\mu\varkappa'_{(ij)} + 2\varepsilon\varkappa'_{\langle ij\rangle} + \beta\varkappa'_{kk}\delta_{ij} .$$

The solution of the system of equation (3.13) satisfies only part of the boundary conditions. If, for example, the boundary is free from loadings, then the condition $\sigma'_{ji}n_j = 0$ is satisfied while condition $\mu'_{ji}n_j = 0$ is not satisfied.

In order to satisfy all the boundary conditions the solution $\underline{u}'', \underline{\zeta}''$ of the system of equations (3.14) is necessary. The boundary conditions connected with the system of equations (3.14) have the form

$$(3.17) \qquad \sigma''_{ji}n_j = 0 \quad , \qquad (\mu'_{ji} + \mu''_{ji})n_j = 0 \quad .$$

E.Betti's theorem of reciprocity of work constitutes one of the most interesting theorems in the theory of elasticity. The theorem is very general and contains the possibility to derive the integration methods of the equations of the theory of elasticity by means of the Green function. We succeeded in generalizing the theorem in the case of micropolar thermoelasticity. For the case of the stationary problem it has the form

$$\int_V (X_i u_i' + Y_i \varphi_i') dV + \int_A (p_i u_i' + m_i \varphi_i') dA + \nu \int_V \vartheta \gamma_{kk}' dV =$$
$$= \int_V (X_i' u_i + Y_i' \varphi_i) dV + \int_A (p_i' u_i + m_i' \varphi_i) dA + \nu \int_V \vartheta' \gamma_{kk} dV. \tag{3.18}$$

We consider here two systems of "generalized forces" acting on an elastic body and the corresponding "generalized displacements". The first group includes the body forces X_i, the body moments Y_i , the tractions p_i, the surface moments m_i , and the temperature field. The displacement $\underline{u}$ the rotation $\underline{\varphi}$ and the temperature ϑ constitute the generalized displacements. The second system of forces and displacements can be distinguished from the first one by the "primes".

Consider a bounded body, rigidly clamped on A_u and free from loadings on A_σ. Assume that the heat sources act in the body, while the surface $A = A_u + A_\sigma$ is heated. We have to determine the displacements u and the rotations φ in the body.

In order to determine the displacements $\underline{u}(\underline{x})$,

$\underline{x} \in V$ we consider a body of the same form and the boundary conditions in the isothermal state. Let the concentrated force $X_i' = \delta(\underline{x} - \underline{\xi})\delta_{ik}$ act at a point $\underline{\xi} \in V$ and produce such a displacement field $U_i^{(k)}(\underline{x},\underline{\xi})$ that it satisfies the homogeneous boundary conditions ($\underline{u}=0$, $\underline{\varphi}=0$ on A_u, $p=0$, $\underline{m}=0$ on A_σ). We obtain from eq. (3.18) for $Y_i'=0$, $Q'=0$

$$(3.19) \qquad u_k(\underline{\xi}) = \gamma \int_V \theta(\underline{x}) \frac{\partial U_i^{k}(\underline{\xi},\underline{x})}{\partial x_i} dV(\underline{x}), \quad \underline{x} \in V, \; k=1,2,3 .$$

The symbol $U_{i,i}^{(k)}(\underline{\xi},\underline{x})$ stands for the dilatation at the point $\underline{\xi}$ due to the action of the concentrated force X_i' situated at the point $\underline{\xi}$.

Let us place at point $\underline{\xi}$ the concentrated moment $Y_i' = \delta(\underline{x} - \underline{\xi})\delta_{ik}$ producing in the body the displacement field $\hat{U}_i^{(k)}(\underline{x},\underline{\xi})$. This field has to satisfy the homogeneous boundary conditions on A_u and A_σ.

Assuming that $X_i'=0$, $Q'=0$, we obtain

$$(3.20) \qquad \varphi_k(\underline{\xi}) = \gamma \int \theta(\underline{x}) \frac{\hat{U}_i^{(k)}(\underline{x},\underline{\xi})}{\partial x_i} dV(\underline{x}), \quad x \in V, \; k=1,2,3 .$$

Eqs. (3.19) and (3.20) constitute a generalization of Maysel's formulae [23] known in the classical thermoelasticity. Eqs. (3.19) and (3.20) are very simple; in order to determine the fields $\underline{u}, \underline{\varphi}$ it is sufficient to integrate eqs. (3.19) and (3.20) provided the Green functions $U_i^{(k)}$, $\hat{U}_i^{(k)}$ have been determined beforehand.

In the particular case of an infinite micropolar body, eqs. (3.19) and (3.20) take the following form

$$u_i(\underline{\xi}) = \frac{\nu}{4\pi(\lambda+2\mu)} \int_V \vartheta(\underline{x}) \frac{\partial}{\partial x_i} \frac{1}{R(\underline{x},\underline{\xi})} dV(\underline{x}), \tag{3.19}$$

$$\varphi_i(\underline{\xi}) = 0 , \tag{3.20}$$

or

$$\left.\begin{aligned} u_i(\underline{\xi}) &= -\frac{m}{4\pi} \int \vartheta(\underline{x}) \frac{\partial}{\partial \xi_i} \left(\frac{1}{R(\underline{\xi},\underline{x})} \right) dV(\underline{x}) , \\ \varphi_i &= 0 . \end{aligned}\right\} \tag{3.21}$$

The above equations are identical with those of the classical thermoelasticity.

Now, consider a single-connected bounded body free from loadings on its surface and free from the body forces and moments inside. The deformations of the body are generated only by its heating. We wish to determine the integrals [24]

$$I_1 = \int_V \operatorname{div} \underline{u}\, dV \quad , \quad I_2 = \int_V \operatorname{div} \underline{\varphi}\, dV , \tag{3.22}$$

characterizing the deformation of the body. The first integral denotes the increase of the body volume, the second one is the mean value of the function $\underline{\varphi} \cdot \underline{n}$ on the body surface. If we assume that the "primed" state corresponds to the uniform tension of the micropolar body $(p_i' = 1 \cdot n_i, \sigma_{ji}' = 1 \cdot \delta_{ij})$ then we obtain from the theorem of reciprocity of works

$$(3.23) \qquad I_1 = 3\alpha_t \int_V \theta(\underline{x})\, dV(\underline{x}) \quad , \qquad \int_V \sigma_{kk}\, dV = 0 \quad .$$

The increase of volume depends here on the temperature distribution in the body, and the mean value of the sum of normal stresses is equal to zero. This result is identical with that of the classical thermoelasticity. Now, assume that the "primed" state corresponds to the uniform torsion $(\mu_{ji}' = 1\,\delta_{ij},\ m_i' = 1\,n_i)$. From the theorem of reciprocity we obtain

$$(3.24) \qquad I_2 = \frac{1}{3\beta + 2\gamma} \int_V \epsilon_{kji} x_k \sigma_{jk}\, dV = \frac{1}{3\beta + 2\gamma} \int_V \mu_{kk}\, dV \, .$$

Let us note that this integral vanishes when the tensor σ_{jk} is symmetric.

So far a number of particular solutions have been obtained. Most of them concern the two-dimensional problems. It is a known fact that two plane strain state problems exist, the first one is characterized by the vectors $\underline{u} = (u_1, u_2, 0)$, $\underline{\varphi} = (0, 0, \varphi_3)$ while the second one by the vectors $\underline{u} = (0, 0, u_3)$ and $\underline{\varphi} = (\varphi_1, \varphi_2, 0)$. Here on the variable x_3. Only the first of these two problems is connected with the field of temperature. A few papers have been devoted to this problem, here we mention the papers by G. Iesan [26], J. Dyszlewicz [27] and W.

Nowacki [28].

Of the two axially symmetric problems, only the first one, characterized by the vectors $\underline{u}=(u_r,0,u_z)$, $\underline{\varphi}=(0,\varphi_\theta,0)$ is connected with the field of temperature. P. Puri [29] and R.S. Dhaliwal [30] have discussed this problem.

Though the main framework of the linear micropolar elasticity has been worked out, a number of particular problems remain unsolved.

The new directions consisting in the incorporation of the further fields are also encouraging. There exists the possibility to construct the theory of thermodiffusion in the micropolar bodies, the coupling of the electrodynamic field with the field of strain, and so on.

An important practical meaning (for the case of action of elevated temperature), may have the theory of non-homogeneous thermoelasticity, which takes into account the material coefficients varying in position and temperature.

REFERENCES

[1] Jeffreys H.: "The thermodynamics of an elastic solid" Proc. Cambridge Phil. Soc., 26, (1930).

[2] Biot M.A.: "Thermoelasticity and irreversible thermodynamics", J. Appl. Phys., 27 (1956).

[3] Parkus H.: "Instationäre Wärmespannungen", Wien (1959).

[4] Boley B.A. and Werner J.H.: "Theory of thermal stresses", John Wiley, New York, (1960).

[5] Nowacki W.: "Thermoelasticity", Pergamon Press, Oxford, (1962).

[6] Cosserat E. and Cosserat F.: "Théorie des corps déformables", A.Herman, Paris, (1909).

[7] Truesdell C. and Toupin R.A.: "The classical field theories" Encycl. of Physics, 3, N°1, Springer Verlag, Berlin (1960).

[8] Grioli G.: "Elasticità asimmetrica", Ann. di Mat. pura ed appl., Ser. IV, 50, (1960).

[9] Günther W.: "Zur Statik und Kinematik des Cosseratschen Kontinuum", Abh. Braunschw. Wiss. Ges. 10, (1958), 85.

[10] Nowacki W.: "Couple-stresses in the theory of thermoelasticity", Proc. of the IUTAM Symposia, Vienna, June 22-28, 1966, Springer Verlag, Wien, (1968).

[11] Nowacki W.: "Couple-stresses in the theory of thermoelasticity" (III), Bull. Acad. Polon. Sci, Sér. Sc. Techn., 14, 505 (1966).

[12] Galerkin B.: "Contributions à la solution générale du problème de la théorie de l'elasticité dans le cas de trois dimensions", C.R. Acad. Sci., Paris, 190, 1047, (1930).

[13] Iacovache M.: "O extindere a metodei lui Galerkin pentru sistemul ecuatiilor elasticitatii", Bull. Stiint. Acad. Rep. Ser. A, 1, (1949), 593.

[14] Sandru N.: "On some problems of the linear theory of symmetric elasticity", Int. J. Engng. Sci. 4, (1966), 81.

[15] Nowacki W.: "On the completeness of potentials in micropolar elasticity", Arch. Mech. Stos. 2, 21, (1969), 107.

[16] Stefaniak J.: "A generalization of Galerkin's functions for asymmetric thermoelasticity", Bull. de l'Acad. Polon. Sci., Sér. Sci. Techn. 16, (1968), 391.

[17] Nowacki W.: "Green functions for micropolar thermoelasticity", Bull. Acad. Polon. Sci., Sér. Sci. Techn. 11-12, 16, (1968).

[18] Iesan D.: "Some theorems in the linear theory of coupled micropolar thermo-viscoelasticity", Bull. de l'Acad. Pol. Sci. Sér. Sci. Techn. 19, 3, (1971), 121.

[19] Kaliski S.: "Thermo-magneto-microelasticity", Bull. de l'Ac. Pol. Sc., Sér. Sc. Techn. 16, 1, (1968), 7.

[20] Nowacki W.: "Variational theorems in asymmetric elasticity", (in Polish), Anniversary book Prof. R. Szewalski, PWN, Warsaw (1968).

[21] Tauchert T.R., Claus W.D. Jr. and Ariman T.: "The linear theory of micropolar thermoelasticity", Int. J. Engng. Sci. 6, 37 (1968).

[22] Schaefer H.: "Das Cosserat-Kontinuum", ZAMM, 47, 8, (1967), 485.

[23] Nowacki W.: "Theory of micropolar elasticity" Springer Verlag, Wien, (1970).

[24] Nowacki W.: "Formulae for overall thermoelastic deformation in a micropolar body", Bull. Acad. Pol. Sc., Sér. Sc. Techn., 1, 17, (1969), 257.

[25] Iesan D.: "On the plane coupled micropolar thermoelasticity", Bull. de l'Ac. Pol. Sci. Sér. Sci. Techn. 1-16, 8, (1968), 379, 2-16, 8 (1968), 385.

[26] Iesan D.: "On thermal stresses in plane strain of isotropic micropolar elastic solids", Acta Mechanica, 11, (1971), 141.

[27] Dyszlewicz J.: "The plane problem of micropolar elasticity", Arch. Mechanics, 24, 1, (1972).

[28] Nowacki W.: "The plane problem of micropolar thermoelasticity", Arch. Mech. Stos. 22, 1, (1970), 3.

[29] Puri P.: "Steady state thermoelastic problem for the half space with couple stresses", Arch. Mech. Stos. 22, 4, (1970), 479.

[30] Dhaliwal R.S.: "The steady-state axisymmetric problem of micropolar thermoelasticity", Arch. Mech. Stos. 23, 5, (1971), 705.

C. WOZNIAK

DISCRETE MICROPOLAR MEDIA

1. Discretized Media

Two well-known approaches are used when we are to discuss the problems of mechanics of solids. One of them is the structural approach, where we take into account the fact that the matter is formed of atoms, and the second one is the continuum approach, where we confine ourselves to the gross phenomena only. However, it is very convenient to introduce also the third approach in which we deal with the model of the solid obtained in the process of discretization, [20]. Such model is said to be the discretized body or the discretized medium and can be applied in many complicated technical problems, where the external loads and properties of the body are described by means of functions having discontinuities or singularities. Theory of discretized bodies and its applications have been lately discussed in a series of papers, [3, 4, 11-18]. In this report we are to give the general review of the theory of discretized bodies in the special case of what is called the discrete elastic micropolar medium.

Because we are not interested in the process of discretization of a continuum, we shall start at once from the definition of a discretized medium. By the discretized medium (or discretized body) we mean the model of a solid given by the pair $(D, \mathcal{E})$, where

1. D is a finite or countable set of elements called particles, $d \in D$, and $\mathcal{E}$ is a covering of D with subsets $E \in D$ called discrete elements, $E \in \mathcal{E}$.

2. Each particle is an independent dynamic holonomic system and all particles have the common configuration space, which is the n-th dimensional vector space V^n .

3. The particles interact in discrete elements, i.e. the forces of interaction among particles in each discrete element are determined by the motion of this discrete element only.

The mechanics of discretized bodies is based on the laws of analytical mechanics and on the principle of determinatism. The general equations of mechanics of discretized bodies have been given in [14] .

In this report we deal only with elastic discretized bodies assuming that for each discrete element there exists the potential which determines the forces of interaction among particles of this discrete element. If each particle is a free material point then $n = 3$ and the configuration space is the physical space. The discretized body is said to be the discrete micropolar medium if $n > 3$, where n is called the number of local degrees of freedom of the medium. Each discretized body is related to the given solid by mens of the constitutive equations; this problem will be discussed in Sec. 3, 4.

To write down explicitly the equations of discretized bodies we have previously to introduce the concept of the allowable difference structure on $(D, \mathcal{E})$. Let us denote $m = \max(\bar{\bar{E}}, \bar{\mathcal{E}}_d) - 1$, where $\mathcal{E}_d := \{E \mid d \in E\}$. The allowable difference structure on $(D, \mathcal{E})$ is a sequence of m one-one mappings $f_\Lambda : D_\Lambda \to D_{-\Lambda}, D_\Lambda \subset D, D_{-\Lambda} \subset D$, satisfying the conditions

$$\bigwedge_\Lambda \bigwedge_{d \in D_\Lambda} f_\Lambda d \neq d, \qquad \bigwedge_{\Lambda,\Phi} \bigwedge_{d \in D_\Lambda \cap D_\Phi} (f_\Lambda d = f_\Phi d) \Rightarrow (\Lambda = \Phi),$$

$$\bigwedge_E \bigvee_{d \in E} E = \{d, f_{\Lambda_1} d, f_{\Lambda_2} d, \dots, f_{\Lambda_s} d\}, \tag{1.1}$$

where $s = s(E) = \bar{\bar{E}} - 1$ (1). Let us assume that the allowable difference structure on $(D, \mathcal{E})$ exists. If such structure is prescribed, then for an arbitrary real-valued function $\varphi : S \to R$, $S \in D$, there exists $2m$ functions $\Delta_\Lambda \varphi : S_\Lambda \to R, \bar{\Delta}_\Lambda \varphi : S_{-\Lambda} \to R$, given by $\Delta_\Lambda \varphi(d) = \varphi(f_\Lambda d) - \varphi(d)$ and $\bar{\Delta}_\Lambda \varphi(d) = \varphi(d) - \varphi(f_{-\Lambda} d)$, respectively, where $S_\Lambda := \{d \mid f_\Lambda d \in S\}$ and $S_{-\Lambda} := \{d \mid f_{-\Lambda} d \in S\}$.
Functions $\Delta_\Lambda \varphi$ and $\bar{\Delta}_\Lambda \varphi$ (some of them may be defined on

(1) Indices $\Lambda, \Phi, \dots$ take the values $I, II, \dots, m$ if otherwise stated. Summation convention holds. The symbols $f_\Lambda d$, $f_{-\Lambda} d$ stand for $f_\Lambda(d)$, $f_{-\Lambda}(d)$, and the symbols $f_{-\Lambda}$, f_{Φ} stand for f_Λ^{-1}, f_Φ^{-1} respectively.

empty sets) are called the first differences of the function φ. Because the differences of an arbitrary real-valued function are real-valued functions, then we can define the second differences of φ as the first differences of first differences: $\Delta_\Lambda \Delta_\Phi \varphi = \Delta_\Lambda(\Delta_\Phi \varphi) : S_{\Phi\Lambda} \to R$. The geometric concepts connected with the theory of discretized bodies are discussed in [17].

2. Equations of Motion

We can prove, [14], that the forces acting at an arbitrary particle $d \in D$ of the discretized body are represented by the vectors in the n-th dimensional vector space V^{*n}, which is dual to the space V^n. Let the allowable difference structure on $(D, \mathcal{E})$ be given in what follows. The forces of interactions in the discrete element E, acting at the particles $d \in E$, $f_\Lambda d \in E$ will be denoted by $-T(d,\tau)$, $-T^\Lambda(d,\tau)$ respectively, where τ is the time coordinate. Moreover, it is very convenient to denote $T^\Lambda(d,\tau) \overset{df}{=} 0$ when $d \sim \in D_\Lambda$ and $T^\Lambda(f_\Lambda d,\tau) \overset{df}{=} 0$ when $d \sim \in D_{-\Lambda}$. The sum of all internal forces in E (i. e. the sum of all forces of interaction among the particles of E) will be defined by the relation

$$t(d,\tau) = -T(d,\tau) - \sum_{\Lambda=1}^{m} T^\Lambda(d,\tau) \tag{2.1}$$

for each $d \in \bigcup D_\Lambda$. if $d \sim \in D_\Lambda$ then we put $t(d,\tau) \overset{df}{=} 0$. Let $f(d,\tau)$ be a vector of external force acting at $d \in D$. The vector of the resultant force $Q(d,\tau)$ acting at d is equal to

$$Q(d,\tau) = -T(d,\tau) - \sum_{\Lambda=1}^{m} T^{\Lambda}(f_{-\Lambda}d,\tau) + f(d,\tau) \; . \qquad (2.2)$$

Lagrange's equations of the second kind of the particle d have the well-known form $Q(d,\tau) = r(d,\ldots)$, where

$$r(d,\ldots) = \frac{d}{d\tau}\frac{\partial T(d,\ldots)}{\partial \dot{q}(d,\tau)} - \frac{\partial T(d,\ldots)}{\partial q(d,\tau)}, \quad q(d,\tau) \in V^n, \qquad (2.3)$$

and $T(d,\ldots) = T(d, q(d,\tau), \dot{q}(d,\tau))$ is the kinetic energy of the particle d (each particle is holonomic dynamic system, cf. Sec. 1). Using (2.2), (2.1) and $Q(d,\tau) = r(d,\ldots)$ we arrive at

$$\bar{\Delta}_\Lambda T^{\Lambda}(d,\tau) + t(d,\tau) + f(d,\tau) = r(d,\ldots), \; d \in D. \qquad (2.4)$$

Equations (2.4) have the same form for all discretized bodies and are said to be the equations of motion of the discretized body.

3. Constitutive Equations

The motion of an arbitrary particle $d \in D$ is given by the vector function $q(d,\tau) \in V$. If the discretized body is elastic, then for each $E \in \mathcal{E}$ there exists the potential $e(E, q(d,\tau), q(f_\Lambda d, \tau))$, where $d \in E$, $f_\Lambda d \in E$, and

$$(3.1) \qquad T(d,\tau) = \frac{\partial e(E,\ldots)}{\partial q(d,\tau)}, \quad T^\Lambda(d,\tau) = \frac{\partial e(E,\ldots)}{\partial q(f_\Lambda d,\tau)}; \quad E \in \mathcal{E}.$$

The motion of E is said to be localized at the particle $d \in E$, if it is given in the form $q(d,\tau) \in \overset{n}{V}, \Delta_\Lambda q(d,\tau) \in \overset{n}{V}$. Let us define the function $\varepsilon(d,\ldots)$, putting

$$(3.2) \qquad \varepsilon(d, q(d,\tau), \Delta_\Lambda q(d,\tau)) \overset{df.}{=} e(E, q(d,\tau), q(d,\tau) + \Delta_\Lambda q(d,\tau)), \qquad E \in \mathcal{E}.$$

The argument d denotes that the motion of E is localized at $d \in E$. By virtue of (3.1), (3.2) and using the relations

$$\frac{\partial e}{\partial q(f_\Lambda d,\tau)} = \frac{\partial \varepsilon}{\partial \Delta_\Lambda q(d,\tau)}, \quad \frac{\partial \varepsilon}{\partial q(d,\tau)} = \frac{\partial e}{\partial q(d,\tau)} + \sum_\Lambda \frac{\partial e}{\partial q(f_\Lambda d,\tau)},$$

we arrive at

$$(3.3) \qquad T^\Lambda(d,\tau) = \frac{\partial \varepsilon(d,\ldots)}{\partial \Delta_\Lambda q(d,\tau)}, \quad t(d,\tau) = -\frac{\partial \varepsilon(d,\ldots)}{\partial q(d,\tau)}; \quad d \in \bigcup D_\Lambda.$$

Eqs. (3.3) are said to be the constitutive equations of hyperelastic discretized bodies, and the function $\varepsilon(d,\ldots)$ is the elastic potential of such body in the discrete element $E=\{d, f_\Lambda d\}$. Equations of motion (2. 4), given for each $d \in D$, and the constitutive equations (3. 3), given for each $E=\{d, f_{\Lambda_1} d, f_{\Lambda_2} d, \dots, f_{\Lambda_\delta} d\}$ form the basic set of equations of what is called the discrete elasticity, [13], or the elasticity of discrete micropolar media. Substituting (3. 3) into (2. 4) we arrive at the set of $n\bar{\bar{D}}$ equations for $n\bar{\bar{D}}$ unknown functions $q^a(d,\tau)$ (each $\mathbf{q}(d,\tau)$ is a sequence of n functions $q^a(d,\tau)$, $a=1,2,\dots,n$). They are ordinary differential equations of the second order with respect to the time coordinate, and have to be considered together with the initial conditions $\mathbf{q}(d,\tau_0)=\mathbf{q}_0(d)$, $\dot{\mathbf{q}}(d,\tau_0)=\mathbf{v}_0(d)$, $d \in D$, where $\mathbf{q}_0(d)$, $\mathbf{v}_0(d)$ are given. In special cases some of the functions $q^a(d,\tau)$, $a=1,2,\dots,n$, may be given a priori; then the suitable functions $f_a(d,\tau)$ have to be considered as unknowns.

The elastic discretized body is related to the given elastic solid by means of the elastic potential $\mathcal{E}(d,\mathbf{q}(d,\tau), \Delta_\Lambda \mathbf{q}(d,\tau))$, given for each $E=\{d, f_{\Lambda_1} d, f_{\Lambda_2} d, \dots, f_{\Lambda_\delta} d\}$. Let us imagine that the elastic continuous body is separated by material surfaces into a set of finite elements. Let us denote by F the finite element of the continuous body B, let $\mathfrak{F}$ be the set of all finite elements, $\chi_F(X,\tau)$, $X \in F$ be the motion of F in the physical space, $\nabla\chi_F(X,\tau)$- the deformation gradient with respect to a given reference configuration of F, and let $\epsilon(X, \nabla\chi_F(X,\tau))$ be a strain energy function with respect to this configuration. Let $(D, \mathcal{E})$ be a given discretized body and let us introduce the one-one mapping $\mu : \mathfrak{F} \to \mathfrak{E}$. We confine our considerations to the motions of B which for each $F \subset B$ can be approximated by the formula

$$(3.4)\qquad \chi_F(X,\tau) = \Phi_F\big(X, q(d,\tau), q(f_\Lambda d,\tau)\big),$$

where $E = \{d, f_{\Lambda_1}d, \ldots, f_{\Lambda_s}d\} = \mu(F)$ and where Φ_F is the known function. The elastic potential of an arbitrary discrete element $E = \mu(F)$ can be now assumed in the form

$$(3.5)\qquad \mathcal{E}\big(d, q(d,\tau), \Delta_\Lambda q(d,\tau)\big) = \int_{\chi(F)} \epsilon\Big(X, \nabla\Phi_F\big(X, q(d,\tau), q(d,\tau) + \Delta_\Lambda q(d,\tau)\big)\Big)\, dF,$$

where $\chi(F)$ is the region in the physical space occupied by the finite element F in the reference configuration. Because the integrands in (3.5) are known functions of $X \in F$, then we can calculate the elastic potential $\mathcal{E}(d,\ldots)$ from (3.5). It follows that each discrete element $E \in \mathfrak{E}$ is a model of a finite element $F = \mu^{-1}(E)$.

4. External Loads and Inertia Forces

Let $b(X,\tau)$, $X \in B$ be the vector density in the physical space describing the external loads acting at the continuous body B. Using the formulae

$$(4.1)\qquad \sum_D f(d,\tau)\,\delta q(d,\tau) = \sum_{F\in\mathfrak{F}} \int_{\chi(F)} b(X,\tau)\cdot\delta\chi_F(X,\tau)\,dF =$$

$$\sum_{F\in\mathfrak{F}} \int_{\chi(F)} b(X,\tau)\cdot \delta\,\Phi_F\big(X, q(d,\tau), q(f_\wedge d,\tau)\big)\,dF$$

we arrive at

$$f(d,\tau) = \sum_{F\in\mathfrak{F}} \int_{\chi(F)} b(X,\tau)\cdot \frac{\partial\,\Phi_F(X,\ldots)}{\partial q(d,\tau)}\,dF. \tag{4.2}$$

The external forces in the discretized body are related to the loads acting at the continuous body by Eqs. (4. 2).

To obtain the kinetic energy $T(d, q(d,\tau), \dot{q}(d,\tau))$ for each particle of the discretized body we have to replace the mass distribution in the continuous body by masses assigned to the particles of the discretized body. In the more general case we can also replace Eq. (2. 3) by the following one

$$r(d,\ldots) = \frac{d}{d\tau}\,\frac{\partial v}{\partial \dot{q}(d,\tau)} - \frac{\partial v}{\partial q(d,\tau)}, \tag{4.3}$$

where

$$v\big(q(d,\tau),\, \dot{q}(d,\tau)\big) = \frac{1}{2}\sum_{F\in\mathfrak{F}} \int_{\chi(F)} P_F(X)\, \dot{\Phi}_F(X,\ldots)\cdot \dot{\Phi}_F(X,\ldots)\,dF \tag{4.4}$$

is the generalized potential of the whole discretized body, and $P_F(X)$, $X \in F$ is the mass distribution in the discrete element $F \in \mathfrak{F}$. This approach was discussed in [13].

Formulas (3. 5), (4. 2) and (4. 4) relate the elastic discretized body (the elastic micropolar medium) to the con tinuous hyperelastic body. We conclude that the equations of the discretized body describe exactly the continuous body subjected to the constraints given by Eqs. (3.4) for each $F \in \mathfrak{F}$.

5. Isotropy

Let $E \in \mathcal{E}$ be a given discrete element and let us denote $s = \bar{\bar{E}}-1$. Each allowable difference structure on $(D, \mathcal{E})$ implies the one-one mapping $\varkappa : E \rightarrow \{0, \Lambda_1, \Lambda_2, \dots, \Lambda_s\}$ " given by $\varkappa(d) = 0$, $\varkappa(f_\Lambda d) = \Lambda$, where $E = \{d, f_{\Lambda_1} d, \dots, f_{\Lambda_s} d\}$ in this difference structure (1). An arbitrary one-one mapping $\varkappa : E \rightarrow \{0, \Lambda_1, \Lambda_2, \dots, \Lambda_s\}$ $s = E-1$ is said to be the coordinate system at E , localised at the point $d = \varkappa^{-1}(0) \in E$.

Let at E be given two coordinate systems $\varkappa : E \rightarrow \{0, \Lambda_1, \dots, \Lambda_s\}$ and $\varkappa' : E \rightarrow \{0, \Lambda'_1, \dots, \Lambda'_s\}$. The one-one mapping $T : \{0, \Lambda_1, \dots, \Lambda_s\} \rightarrow \{0, \Lambda'_1, \dots, \Lambda'_s\}$ given by $T = \varkappa' \circ \varkappa^{-1}$ is said to be a trasnformation of the coordinate system; all such transformations form a group. To each transformation $\varkappa$ we can assign $(s+1) \times (s+1)$ permutation matrix

$$(5.1) \quad (A_{\lambda'}{}^{\lambda}) = \begin{pmatrix} A_0{}^0 & A_0{}^\Lambda \\ A_{\Lambda'}{}^0 & A_{\Lambda'}{}^\Lambda \end{pmatrix}; \quad \lambda = 0, \Lambda_1, \dots, \Lambda_s \quad ; \quad \lambda' = 0, \Lambda'_1, \dots, \Lambda'_s ,$$

(1) In this section the index Λ takes the values $\Lambda_1, \Lambda_2, \dots, \Lambda_s$, and the index Λ' run over the sequences $\Lambda'_1, \Lambda'_2, \dots, \Lambda'_s$, where $\Lambda_1 < \Lambda_2 < \dots \Lambda_s$ and $\Lambda'_1 < \Lambda'_2 < \dots \Lambda'_s$. Summation convention for Λ holds.

putting $A_{\lambda'}^{\lambda} = 1$ when $\lambda' = T(\lambda)$ and $A_{\lambda'}^{\lambda} = 0$ when $\lambda' \neq T(\lambda)$. Now we can say that at the discrete element E there is defined an object with k components if to each coordinate system $\varkappa$ at E a sequence of k members $\omega_1,\ldots,\omega_k$ has been assigned. The object is called geometric id its components $\omega'_1,\ldots,\omega'_k$ in an arbitrary coordinate system $\varkappa'$ can be expressed by the components $\omega_1,\ldots,\omega_k$ and by the transformation $T = \varkappa' \circ \varkappa^{-1}$. To give a simple example of a geometric object let us introduce the function $\varphi : E \to R$. Denoting $\varphi_0 = \varphi(\varkappa^{-1}(0))$, $\Delta_\Lambda \varphi = \varphi(\varkappa^{-1}(\Lambda)) - \varphi(\varkappa^{-1}(0))$ and $\varphi'_0 = \varphi(\varkappa'^{-1}(0))$, $\Delta_{\Lambda'}\varphi = \varphi(\varkappa'^{-1}(\Lambda)) - \varphi(\varkappa'(0))$,we can prove that the following relation holds

$$\begin{pmatrix} \varphi'_0 \\ \Delta_{\Lambda'}\varphi \end{pmatrix} = \begin{pmatrix} 1 & A_0^{\Lambda} \\ 0 & B_{\Lambda'}^{\Lambda} \end{pmatrix} \begin{pmatrix} \varphi_0 \\ \Delta_\Lambda \varphi \end{pmatrix}, \tag{5.2}$$

where $B_{\Lambda'}^{\Lambda} = A_{\Lambda'}^{\Lambda} - I_{\Lambda'} A_0^{\Lambda}$ and $I_{\Lambda'} = 1$ for each Λ' . The $s+1$ numbers φ_0, $\Lambda_\Lambda \varphi$ are components of the certain geometric object at E . The set of all $(s+1)\times(s+1)$ matrices (5.1) forms the group which will be denoted by $\mathfrak{U}^{s+1}$ and which is a representation of the transformation group. For further discussion of the geometric object the reader is referred to [14] .

The form of the elastic potential at the discrete element E depends on the choice of the coordinate system at E . Using the formula (5. 2) we obtain

$$\varepsilon\big(d, \mathbf{q}(d,\tau), \Delta_\Lambda \mathbf{q}(d,\tau)\big) = \varepsilon'\big(d', \mathbf{q}(d',\tau), \Delta_{\Lambda'}\mathbf{q}(d',\tau)\big) = \tag{5.3}$$

$$= \varepsilon'\left(d, q(d,\tau) + A_0^{\Lambda} \Delta_{\Lambda} q(d,\tau), B_{\Lambda'}^{\Lambda} \Delta_{\Lambda} q(d,\tau)\right)$$

It is easy to prove that there always exists the subgroup $\mathfrak{J}^{s+1}$ of the group $\mathfrak{U}^{s+1}$, such that the functions $\varepsilon, \varepsilon'$ are identical, i.e. such that the relation

(5.4)

$$\varepsilon\left(d, q(d,\tau), \Delta_{\Lambda} q(d,\tau)\right) = \varepsilon\left(d, q(d,\tau) + A_0^{\Lambda} \Delta_{\Lambda} q(d,\tau), B_{\Lambda'}^{\Lambda} \Delta_{\Lambda} q(d,\tau)\right).$$

holds for each matrix $\left(A_{\lambda'}^{\lambda}\right) \in \mathfrak{J}^{s+1}$. The group $\mathfrak{J}^{s+1}$, which may be trivial (then $\left(A_{\lambda'}^{\lambda}\right)$ is a unit matrix) is said to be the isotropy group of the elastic potential at E. From (5.4) follows that the elastic potential does not change if we replace the motion $q(d,\tau), q(f_{\Lambda_1}, d, \tau), \ldots, q(f_{\Lambda_s}, d)$ of E by the motion $q(d',\tau), q(f_{\Lambda'_1}, d', \tau), \ldots, q(f_{\Lambda_s}, d, \tau)$ respectively. If $\mathfrak{J}^{s+1} = \mathfrak{U}^{s+1}$ then the elastic discrete element is said to be isotropic.

6. Small Motions Superimposed on Arbitrary Motion

Let us introduce in the vector space V^n the vector basis. The constitutive equations (3.3) have then the form

$$(6.1) \quad T_a^{\Lambda}(d,\tau) = \frac{\partial \varepsilon(d,\ldots)}{\partial \Delta_{\Lambda} q^a(d,\tau)}, \quad t_a(d,\tau) = -\frac{\partial \varepsilon(d,\ldots)}{\partial q^a(d,\tau)}; \quad a = 1, 2, \ldots, n,$$

where $\mathcal{E}(d,\ldots) = \mathcal{E}\left(d, q^a(d,\tau), \Delta_\Lambda q^a(d,\tau)\right)$, and where $q^a(d,\tau)$ are generalized coordinates and $T_a^\Lambda(d,\tau)$ are generalized forces (1). From (2.4) we arrive at the following form of the equations of motion

$$\bar{\Delta}_\Lambda T_a^\Lambda(d,\tau) + t_a(d,\tau) + f_a(d,\tau) = r_a(d,\ldots). \tag{6.2}$$

The set of functions $q^a(d,\tau)$, $d \in D$, satisfying (6.1) and (6.2) as well as the prescribed initial conditions is said to be the fundamental motion.

Now we are to study the motion

$$^*q^a(d,\tau) = q^a(d,\tau) + \epsilon\, w^a(d,\tau), \qquad d \in D, \tag{6.3}$$

where ϵ is the small parameter (the squares and the high powers of ϵ will be neglected compared with ϵ). The set of functions $w^a(d,\tau)$, $d \in D$, is said to be the superposed motion, which is assumed to be independent of the fundamental motion. Denoting by $^*H$ an arbitrary quantity relating to the motion (6.3) we can write

$$^*H = H + \epsilon\, 'H \tag{6.4}$$

where H relates to the fundamental motion. By virtue of (6.4) we obtain

$$'\mathcal{E} = \frac{\partial \mathcal{E}}{\partial q^a} w^a + \frac{\partial \mathcal{E}}{\partial \Delta_\Lambda q^a} \Delta_\Lambda w^a, \quad 'T = \frac{\partial T}{\partial q^a} w^a + \frac{\partial T}{\partial \dot{q}^a} \dot{w}^a, \tag{6.5}$$

(1) Indices $a, b, \ldots$ take the values $1, 2, \ldots n$; summation convention holds.

$$'T_a^\Lambda = \frac{\partial' \varepsilon}{\partial \Delta_\Lambda q^a}, \qquad 't_a = -\frac{\partial' \varepsilon}{\partial q^a}.$$

After denotations

$$(6.6) \quad K_{ab}^{\Lambda\Phi} = \frac{\partial^2 \varepsilon}{\partial \Delta_\Lambda q^a \partial \Delta_\Phi q^b} \quad L_{ab}^\Lambda = \frac{\partial^2 \varepsilon}{\partial \Delta_\Lambda q^a \partial q^b}, \quad M_{ab} = \frac{\partial^2 \varepsilon}{\partial q^a \partial q^b},$$

we arrive at the fomulas

$$(6.7) \quad \begin{aligned} 'T_a^\Lambda &= K_{ab}^{\Lambda\Phi} \Delta_\Phi w^b + L_{ab}^\Lambda w^b, \\ 't_a &= -L_{ba}^\Phi \Delta_\Phi w^b - M_{ab} w^b, \end{aligned}$$

which are said to be the constitutive equations for the superimposed motion. $w^a(d,\tau)$, $d \in D$. Using (6.4) and (6.5) we also obtain

$$(6.8) \quad \bar{\Delta}_\Lambda 'T_a^\Lambda + 't_a + 'f_a = 'r_a,$$

where

$$(6.9) \quad 'r_a = \frac{d}{d\tau} \frac{\partial' T}{\partial \dot{q}^a} - \frac{\partial' T}{\partial q^a}.$$

Eqs. (6.8) are said to be the equations of motion for the superimposed motion. If the fundamental motion is known, then from (6.7), (6.8) and from the initial conditions for the superimposed motion we can determine the functions $w^a(d,\tau)$, $d \in D$.

Now, we are able to formulate the problem of stability in the mechanics of elastic discreteized bodies. Let us suppose that the fundamental motion reduces to an equilibri um state, i.e. $\dot{q}^a(d,\tau) = 0$ for each $d \in D$ and τ. Moreover, let each particle $d \in D$ be the scleronomic dynamical system. In this case

$$T = \frac{1}{2} a_{ab} \dot{q}^a \dot{q}^b , \quad 'T = \frac{1}{2} a_{ab} (\dot{w}^a \dot{q}^b + \dot{q}^a \dot{w}^b),$$

where $a_{ab}(d)$ are constant for each $d \in D$. By virtue of (6.7) and (6.8) we arrive at

$$\bar{\Delta}_\Lambda (K^{\Lambda\Phi}_{ab} \Delta_\Phi w^b + L^\Lambda_{ab} w^b) - L^\Lambda_{ab} \Delta_\Lambda w^b - M_{ab} w^b = a_{ab} \ddot{w}^b, \quad (6.10)$$

An equilibrium state described by the functions $q^a(d)$ is said to be stable if the amplitude of the superimposed motion is always vanishingly small when the disturbance itself is sufficiently small. Substituting $w^a = u^a(d) e^{i\tilde{\omega}\tau}$ to (6.10) we obtain

$$\bar{\Delta}_\Lambda (K^{\Lambda\Phi}_{ab} \Delta_\Phi u^b + L^\Lambda_{ab} u^b) - L^\Lambda_{ba} \Delta_\Lambda u^a - M_{ab} u^b = - \tilde{\omega}^2 a_{ab} u^b, \quad (6.11)$$

The equilibrium state $q(d)$, $d \in D$ is stable if small oscillations $w^a = u^a(d) e^{i\tilde{\omega}\tau}$ are limited for each τ. It follows that the general criterion of stability of the discretized body has the form $\Im m\, \tilde{\omega} \geq 0$. Using the statical criterion we assume that $\tilde{\omega} = 0$ in (6.11); if there exists only the trivial solution of (6.11), then the state $q(d)$, $d \in D$ is stable. Problems of superimposed motions were studied in [18].

7. General Coordinates

In many special problems it is very convenient to assign the separate vector basis in V^n for each $d \in D$. Such bundle of vector bases in V^n is said to be the general coordinates in V^n. The components of an arbitrary vector $u(d) \in V^n$ in general coordinates will be given by $u^\alpha(d) = A^\alpha_a(d)\, u^a(d)$, where $(A^\alpha_a(d))$ is $n \times n$ non-singular matrix, given for each $d \in D$ and indices $\alpha, \beta, \ldots$ take the values $1, 2, \ldots, n$. The components of an arbitrary covector in V^{*n} are given by $v_\alpha(d) = A^a_\alpha(d)\, v_a(d)$ where $A^\alpha_a(d) A^b_\alpha(d) = \delta^b_a$. The vector with components

(7.1)
$$\delta_\Lambda u^\alpha(d) = A^\alpha_a(d)\, \Delta_\Lambda\left[A^a_\beta(d)\, u^\beta(d)\right] = \Delta_\Lambda u^\alpha(d) + G^\alpha_{\Lambda\beta}(d)\, u^\beta(f_\Lambda d), \quad d \in D_\Lambda,$$

is called the absolute difference of the vector field $u(d) \in V^n$, $d \in D$, and the covector with components

(7.2)
$$\bar{\delta}_\Lambda v_\alpha(d) = A^a_\alpha(d)\, \bar{\Delta}_\Lambda\left[A^\beta_a(d)\, v_\beta(d)\right] = \bar{\Delta}_\Lambda v_\alpha(d) + \overset{*}{G}{}^\beta_{\alpha\Lambda}(d)\, v_\beta(f_{\Lambda} d), \quad d \in D_{-\Lambda},$$

is said to be the absolute difference of the covector field $v(d) \in V^{*n}$, $d \in D$; from (7.1) and (7.2) follows that

$$(7.3) \qquad G^\alpha_{\Lambda\beta}(d) = A^\alpha_a(d)\, \Delta_\Lambda A^a_\beta(d) \qquad \overset{*}{G}{}^\beta_{a\Lambda}(d) = A^a_\alpha(d)\, \bar{\Delta}_\Lambda A^\beta_a(d) .$$

The matrices (7.3) are called the connection matrices. The general theory of connection in the bundle of vector spaces over D has been given in [17].

Using the matrices $(A^\alpha_a(d))$, $(A^a_\alpha(d))$ we can easily transform the equations to the general coordinates. From eqs. (6.1) we obtain the following form of the constitutive equations

$$T_\alpha^\Lambda(d,\tau) = \frac{\partial \varepsilon(d,\ldots)}{\partial \delta_\Lambda q^\alpha(d,\tau)}, \quad t_\alpha(d,\tau) = -\frac{\partial \varepsilon(d,\ldots)}{\partial q^\alpha(d,\tau)}, \tag{7.4}$$

where $\varepsilon(d,\ldots) = \varepsilon\left(d, q^\alpha(d,\tau), \delta_\Lambda q^\alpha(d,\tau)\right)$. The equations of motion (6.2) have the form

$$\bar{\delta}_\Lambda T_\alpha^\Lambda(d,\tau) + t_\alpha(d,\tau) + f_\alpha(d,\tau) = r_\alpha(d,\ldots), \tag{7.5}$$

where

$$r_\alpha(d,\ldots) = \frac{d}{d\tau}\,\frac{\partial T(d,\ldots)}{\partial \dot{q}^\alpha(d,\tau)} - \frac{\partial T(d,\ldots)}{\partial q^\alpha(d,\tau)} \tag{7.6}$$

and $T(d,\ldots) = T(d, q^\alpha(d,\tau), \dot{q}^\alpha(d,\tau)$. For the superimposed motion we obtain from (6.7) the constitutive equations

$$\begin{aligned} {}'T_\alpha^\Lambda &= K_{\alpha\beta}^{\Lambda\Phi}\, \delta_\Phi w^\beta + L_{\alpha\beta}^\Lambda\, w^\beta , \\ {}'t_\alpha &= -L_{\beta\alpha}^\Phi\, \delta_\Phi w^\beta - M_{\alpha\beta}\, w^\beta , \end{aligned} \tag{7.7}$$

and from (6.8) the equations of motion in general coordinates

$$\bar{\delta}\, {}'T_\alpha^\Lambda + {}'t_\alpha + {}'f_\alpha = {}'r_\alpha . \tag{7.8}$$

The equations of discretized bodies in general coordinates were introduced for the first time in [13] . This concept has also been studied in [1, 2] .

8. Linear and Second Order Theories

Let N, $0 < N \leqslant n$ be a given positive integer, and let the indices $K, L, \ldots = 1, 2, \ldots, N-1$ stand for $\alpha, \beta, \ldots$ when $\alpha, \beta, \ldots < N$, and let the indices $A, B, \ldots = N, N+1, \ldots, n$ stand for $\alpha, \beta, \ldots$ when $N \leqslant \alpha, \beta, \ldots \leqslant n$. Indices $K, L, \ldots$ and $A, B, \ldots$ are related to general coordinates in V^n, where $q^\alpha(d,\tau) = 0$ refer to the stable state of equilibrium and $\varepsilon(d, \ldots) = 0$, $\partial\varepsilon(d,\ldots)/\partial q^\alpha(d,\tau) = 0$ $\partial\varepsilon(d,\ldots)/\partial\delta_\Lambda q^\alpha(d,\tau) = 0$ for $q^\alpha(d,\tau) = 0$ and each $d \in D$.

Let the fundamental motion be given by $q^\alpha(d,\tau) = \delta^\alpha_k q^k(d,\tau)$ where the deviations $q^k(d,\tau)$ from the state of equilibrium and the generalized velocities $\dot{q}^k(d,\tau)$ are sufficiently small in absolute value. This motion will be governed by the following linearized equations

$$
\begin{aligned}
&\bar{\delta}_\Lambda T_k^\Lambda + t_k + f_k = r_k; \\
&T_k^\Lambda = \overset{\circ}{K}{}^{\Lambda\Phi}_{kL}\,\delta_\Phi q^L + \overset{\circ}{L}{}^{\Lambda}_{kL} q^L, \quad t_k = -\overset{\circ}{L}{}^{\Phi}_{Lk}\,\delta_\Phi q^L - \overset{\circ}{M}_{kL} q^L,
\end{aligned}
\tag{8.1}
$$

where we have denoted

$$
\overset{\circ}{K}{}^{\Lambda\Phi}_{kL} = \left.\frac{\partial^2\varepsilon}{\partial\delta_\Lambda q^k\,\partial\delta_\Phi q^L}\right|_{q=0}, \quad \overset{\circ}{L}{}^{\Phi}_{kL} = \left.\frac{\partial^2\varepsilon}{\partial\delta_\Phi q^k\,\partial q^L}\right|_{q=0}, \quad \overset{\circ}{M}_{kL} = \left.\frac{\partial^2\varepsilon}{\partial q^k\,\partial q^L}\right|_{q=0},
\tag{8.2}
$$

and $q = 0$ denotes the state $q^\alpha(d,\tau) = 0$ for each $d \in D$.

Now suppose the superimposed motion in the form $w^\alpha(d,\tau) = \delta^\alpha_\Lambda w^\Lambda(d,\tau)$. Equations of this motion can also be linearized with respect to $q^k(d,\tau)$, and from (7.7), (7.8) and by virtue of

$$K^{\Lambda\Phi}_{\alpha\beta} = \frac{\partial^2 \varepsilon}{\partial\delta_\Lambda q^\alpha \partial\delta_\Phi q^\beta}, \quad L^{\Phi}_{\alpha\beta} = \frac{\partial^2 \varepsilon}{\partial\delta_\Phi q^\alpha \partial q^\beta}, \quad M_{\alpha\beta} = \frac{\partial^2 \varepsilon}{\partial q^\alpha \partial q^\beta},$$

we obtain

$$\bar{\delta}_\Lambda {}'T^\Lambda_A + {}'t_A + {}'f_A = {}'r_A ; \tag{8.3}$$

$$'T^\Lambda_A = \left(\overset{\circ}{K}{}^{\Lambda\Phi}_{AB} + \overset{\circ}{K}{}^{\Lambda\Phi\Delta}_{ABK}\delta_\Delta q^k + \overset{\circ}{K}{}^{\Lambda\Phi}_{ABK} q^k\right)\delta_\Phi w^B + \left(\overset{\circ}{L}{}^{\Lambda}_{AB} + \overset{\circ}{L}{}^{\Lambda\Phi}_{ABK}\delta_\Phi q^k + \overset{\circ}{L}{}^{\Lambda}_{ABK} q^k\right) w^B,$$

$$'t_A = -\left(\overset{\circ}{L}{}^{\Phi}_{AB} + \overset{\circ}{L}{}^{\Lambda\Phi\Delta}_{BAK}\delta_\Delta q^k + \overset{\circ}{L}{}^{\Phi}_{BAK} q^k\right)\delta_\Phi w^B + \overset{\circ}{M}_{AB} + \overset{\circ}{M}{}^{\Phi}_{ABK}\delta_\Phi q^k + \overset{\circ}{M}_{ABK} q^k\big) w^B,$$

where (8.4)

$$\overset{\circ}{K}{}^{\Lambda\Phi}_{AB} = \left.\frac{\partial^2 \varepsilon}{\partial\delta_\Lambda q^A \partial\delta_\Phi q^B}\right|_{q=0}, \quad \overset{\circ}{K}{}^{\Lambda\Phi\Delta}_{ABK} = \left.\frac{\partial^3 \varepsilon}{\partial\delta_\Lambda q^A \partial_\Phi q^B \partial_\Delta q^K}\right|_{q=0}, \ldots, \quad \overset{\circ}{M}_{ABK} = \left.\frac{\partial^3 \varepsilon}{\partial q^A \partial q^B \partial q^K}\right|_{q=0}.$$

The superimposed motion $w^\alpha = \delta^\alpha_A w^A(d,\tau)$ is possible only if $\delta_\Lambda {}'T^\Lambda_k + {}'t_k + {}'f_k = {}'r_k$, and the fundamental motion $q^\alpha = \delta^\alpha_k q^k(d,\tau)$ is possible only if $\delta_\Lambda T^\Lambda_A + t_A + f_A = r_A$.

The theory described by Eqs. (8.1) and (8.3) is said to be the second order theory. The characteristic feature of this theory is that the fundamental motion is governed by the linear equations (8.1). By virtue of (8.2) and (8.4) we also obtain

$$\varepsilon = \frac{1}{2}\overset{\circ}{K}{}^{\Lambda\Phi}_{KL}\delta_\Lambda q^K \delta_\Phi q^L + \overset{\circ}{L}{}^{\Lambda}_{KL}\delta_\Lambda q^K q^L + \frac{1}{2}\overset{\circ}{M}_{KL} q^K q^L + \frac{1}{2}\big(\overset{\circ}{K}{}^{\Lambda\Phi}_{AB} +$$

$$+ \overset{\circ}{K}{}^{\Lambda\Phi\Delta}_{ABK} \delta_\Delta q^K + \overset{\circ}{K}{}^{\Lambda\Phi}_{ABK} q^K \Big) \delta_\Lambda q^A \delta_\Phi q^B + \Big(\overset{\circ}{L}{}^{\Lambda}_{AB} + L^{\Lambda\Delta}_{ABK} \delta_\Delta q^K +$$

$$+ \overset{\circ}{L}{}^{\Lambda}_{ABK} q^K \Big) \delta_\Lambda q^A q^B + \frac{1}{2} \Big(\overset{\circ}{M}_{AB} + \overset{\circ}{M}{}^{\Delta}_{ABK} \delta_\Delta q^K + \overset{\bullet}{M}_{ABK} q^K \Big) q^A q^B , \tag{8.5}$$

where $\overset{\circ}{K}{}^{\Lambda\Phi}_{KL}$, $\overset{\circ}{L}{}^{\Lambda}_{KL}, \ldots, \overset{\circ}{M}_{ABK}$ are independent of the motion. The stability problems in the second order theory are formulated in the same way as in the general case (cf. Sec. 6), but there are many problems governed by the linearized equations which can not be solved with the aid of the second order theory.

In the special case $N = n$ we arrive at the linear theory of elastic discretized bodies (linear discrete elasticity). The equations of motion and the constitutive equations of the linear theory are given by

$$\bar{\delta}_\Lambda T_\alpha + t_\alpha + f_\alpha + r_\alpha$$

$$T^\Lambda_\alpha = K^{\Lambda\Phi}_{\alpha\beta} \delta_\Phi q^\beta + L^\Lambda_{\alpha\beta} q^\beta , \quad t_\alpha = - L^\Lambda_{\beta\alpha} \delta_\Lambda q^\beta - M_{\alpha\beta} q^\beta ; \tag{8.6}$$

$$r_\alpha = a_{\alpha\beta} \ddot{q}^\beta ,$$

where

$$\varepsilon = \frac{1}{2} K^{\Lambda\Phi}_{\alpha\beta} \delta_\Lambda q^\alpha \delta_\Phi q^\beta + L^\Lambda_{\alpha\beta} \delta_\Lambda q^\alpha q^\beta + \frac{1}{2} M_{\alpha\beta} q^\alpha q^\beta \tag{8.7}$$

is the elastic potential.

9. Laws of Conservation

Let us denote by x^k, $k = 1,2,3$, the inertial Cartesian coordinates in the physical space. The infinitesimal translation and rotation of the physical space is given by $x^k \to x^k + \epsilon^k + \epsilon^{k\ell} x_\ell$, where $\epsilon^k, \epsilon^{k\ell} = -\epsilon^{\ell k}$, are arbitrary infinitesimal constants. Let us assume that the variations of the functions $q^a(d,\tau)$, $d \in D$ due to the translations and rotations of the physical space, are determined by the formulas

$$q^a(d,\tau) \to q^a(d,\tau) + C^a_k \epsilon^k + C^a_{bk\ell}\epsilon^{k\ell} q^b(d,\tau), \tag{9.1}$$

where C^a_k, $C^a_{bk\ell} = -C^a_{b\ell k}$ are constants and $C^a_{bk\ell} = 0$ for $a \neq b$. Assuming that the elastic potential and the kinetic energy are invariant under arbitrary translations and rotations of the physical space, we can write

$$C^a_k \frac{\partial \varepsilon}{\partial q^a} = 0, \qquad C^a_k \frac{\partial T}{\partial q^a} = 0;$$

$$C^a_{bk\ell}\left(\frac{\partial \varepsilon}{\partial q^a} q^b + \frac{\partial \varepsilon}{\partial \Delta_\Lambda q^a}\Delta_\Lambda q^b\right) = 0, \quad C^a_{bk\ell}\left(\frac{\partial T}{\partial q^a} q^b + \frac{\partial T}{\partial \dot{q}^a}\dot{q}^b\right) = 0. \tag{9.2}$$

We also assume that the functions $\varepsilon(d,\ldots)$ and $T(d,\ldots)$ are invariant under arbitrary "translations" of the inertial time coordinate, given by $\tau \to \tau + \epsilon$ where ϵ is an arbitrary infinitesimal constant. It follows that

$$(9.3)\qquad \frac{\partial \mathcal{E}}{\partial \tau} = 0\,, \qquad \frac{\partial T}{\partial \tau} = 0\,.$$

Equations (9.2) and (9.3) are conditions of existence of conservation laws in the theory of elastic discretized bodies. By virtue of (9.2), (9.3) and using the equations of motion in the form

$$\frac{\partial \mathcal{E}}{\partial q^a} = \bar{\Delta}_\Lambda T_a^{\ \Lambda} + f_a - \frac{d}{d\tau}\frac{\partial T}{\partial \dot{q}^a} + \frac{\partial T}{\partial q^a}$$

we arrive at

$$\frac{d}{d\tau}\left(C_k^a \frac{\partial T}{\partial \dot{q}^a}\right) = C_k^a \bar{\Delta}_\Lambda T_a^{\ \Lambda} + C_k^a f_a\,,$$

$$(9.4)\qquad \frac{d}{d\tau}\left(C_{bkl}^a \frac{\partial T}{\partial \dot{q}^a} q^b\right) = C_{bkl}^a\left(\bar{\Delta}_\Lambda T_a^{\ \Lambda} q^b + T_a^{\Lambda}\Delta_\Lambda q^b\right) + C_{bkl}^a f_a q^b\,,$$

$$\frac{d}{d\tau}\left(\mathcal{E} + T\right) = \bar{\Delta}_\Lambda T_a^{\ \Lambda}\dot{q}^a + T_a^{\Lambda}\Delta_\Lambda \dot{q}^a + f_a \dot{q}^a\,; \qquad d \in D\,.$$

In $(9.4)_3$ we have assumed that the particles $d \in D$ are scleronomic holonomic dynamic systems i. e. the relation

$$2T = \frac{\partial T}{\partial \dot{q}^a}\dot{q}^a$$

holds for each $d \in D$. The expressions in parenthesis on the left-hand sides of $(9.4)_1$ and $(9.4)_2$ are said to be a momentum and the moment of momentum of the particle $d \in D$, respectively. The expression in parenthesis on the left-hand side of $(9.4)_3$ is the sum of the kinetic energy of the particle $d \in D$ and the elastic potential of the discrete element $E = \{d, f_{\Lambda_1}d, \ldots, f_{\Lambda_s}d\}$. Eqs. (9.4) represent the local form of the conservation laws. Eqs. $(9.2)_1$ and $(9.2)_3$ which may be written in the form

$$C_k^a t_a = 0, \quad C_{bkl}^a \left(T_a^\Lambda \Delta_\Lambda q^b - t_a q^b\right) = 0, \tag{9.5}$$

represent the fact, that the resultant force and the resultant moment of internal forces in the discrete element are equal to zero.

To obtain the global form of the conservation laws we have to introduce some auxiliary concepts. Let S be an arbitrary subset of D and let us define the subset $\partial_\Lambda S$ putting $\partial_\Lambda S = (S - S_\Lambda) \cup (S - S_{-\Lambda})$. The subset $\partial_\Lambda S$ is said to be the Λ -boundary of S. It is easy to verify that for each $S \subset D$ we have

$$\sum_{S_\Lambda \cap S_{-\Lambda}} \Delta_\Lambda \varphi(d) = \sum_{\partial_\Lambda S_{-\Lambda}} \varphi(d) N_\Lambda(d), \qquad N_\Lambda(d) \begin{cases} 1 & d \in S_\Lambda - S_{-\Lambda\Lambda} \\ -1 & d \in S_{-\Lambda} - S_{\Lambda-\Lambda} \end{cases} \tag{9.6}$$

where $\varphi : S_{-\Lambda} \to R$ is an arbitrary function. Denoting

$$\sum_{S_0} = \sum_{\Lambda} \sum_{S_\Lambda \cap S_{-\Lambda}}, \qquad \sum_{\partial S} = \sum_{\Lambda} \sum_{\partial_\Lambda S_{-\Lambda}} \tag{9.7}$$

we can also prove that the following finite divergence theorem holds

$$(9.8)\qquad \sum_{S_0} \Delta_\Lambda \varphi^\Lambda(d) = \sum_{\partial S} \varphi^\Lambda(d) N_\Lambda(d),$$

where $\varphi^\Lambda(d)$ is defined on S_Λ . Denoting $\bar{\varphi}^\Lambda(d) \overset{df}{=} \varphi^\Lambda(\rho_{-\Lambda} d)$ we obtain

$$(9.9)\qquad \bar{\Delta}_\Lambda \varphi^\Lambda = \Delta_\Lambda \bar{\varphi}^\Lambda, \qquad \zeta \bar{\Delta}_\Lambda \varphi^\Lambda + \varphi^\Lambda \Delta_\Lambda \zeta = \Delta_\Lambda(\zeta \bar{\varphi}^\Lambda),$$

where $\zeta: S \to R$ is an arbitrary function.

Using (9. 8) and (9. 9) we obtain from (9. 4)

$$\frac{d}{d\tau} \sum_{S_0} \left(C_k^a \frac{\partial T}{\partial \dot{q}^a} \right) = \sum_{\partial S} C_k^a T_a^{(N)} + \sum_{S_0} C_k^a f_a ,$$

$$(9.10)\qquad \frac{d}{d\tau} \sum_{S_0} \left(C_{bkl}^a \frac{\partial T}{\partial \dot{q}^a} q^b \right) = \sum_{\partial S} C_{bkl}^a T_a^{(N)} q^b + \sum_{S_0} C_{bkl}^a f_a q^b ,$$

$$\frac{d}{d\tau} \sum_{S_0} (\varepsilon + T) = \sum_{\partial S} T_a^{(N)} \dot{q}^a + \sum_{S_0} f_a \dot{q}^a ,$$

where we have denoted

$$(9.11)\qquad T_a^{(N)} = \bar{T}_a^\Lambda N_\Lambda .$$

Equations (9. 10) are the global form of the conservation laws in the mechanics of elastic discretized bodies.

10. Stresses and Strains

Let the elastic potential $\varepsilon(d,\ldots)$ be invariant, for each $d \in U D_\Lambda$, only under the group (9.1) of infinitesimal transformations. Let us also assume that the general solution of Eqs. $(9.2)_1$ and $(9.2)_3$ can be given in the form

$$\varepsilon = \varepsilon\left(d, \gamma_A(d, \tau)\right), \tag{10.1}$$

where

$$\gamma_A(d,\tau) = \varphi_A\left(d, \overset{a}{q}(d,\tau), \Delta_\Lambda(d,\tau)\right), \qquad A = 1,2,\ldots K, \tag{10.2}$$

are differentiable functions and K is a constant for each $d \in U D_\Lambda$. The values of the functions (10.2) are said to be the discrete strains at the given discrete element $E = \{d, f_{\Lambda_1}d, \ldots, f_{\Lambda_s}d\}$. We can also introduce the following functions

$$\overset{A}{p}(d,\tau) = \frac{\partial\varepsilon\left(d, \gamma_B(d,\tau)\right)}{\partial\gamma_A(d,\tau)}, \tag{10.3}$$

which are said to be the discrete stresses in the discrete element $E = \{d, f_{\Lambda_1}d, \ldots, f_{\Lambda_s}d\}$. Because of

$$\overset{\Lambda}{T_a} = \overset{\Lambda}{\Phi}_{Aa}\overset{A}{p}, \qquad t_a = -\overset{A}{p}\Phi_{Aa}, \tag{10.4}$$

where we have denoted

$$\overset{\Lambda}{\Phi}_{Aa} = \frac{\partial\varphi_A\left(d, \overset{b}{q}, \Delta_\Phi \overset{b}{q}\right)}{\partial\Delta_\Lambda \overset{a}{q}}, \qquad \Phi_{Aa} = \frac{\partial\varphi_A\left(d, \overset{b}{q}, \Delta_\Phi \overset{b}{q}\right)}{\partial \overset{a}{q}}, \tag{10.5}$$

the equations of motion (6. 2) can be written in the form

$$\bar{\Delta}_A\left(\Phi^A_{Aa} p^A\right) + \Phi_{Aa} p^A + f_a = r_a . \tag{10.6}$$

Equations of motion (10. 6), constitutive equations in the form (10. 3) and "geometric" equations (10. 2) form the alternative system of equations of the discretized elastic body. In many special cases we can assume the elastic potential in the form

$$\mathcal{E} = \frac{1}{2} A^{AB} \gamma_A \gamma_B . \tag{10.7}$$

The constitutive equations (10. 3) are then linear with respect to the strains γ_A and we arrive at what is called the small deformation theory of discretized bodies.

The more general approach to the problem of strains and stresses in discretized bodies, in which the discrete elements are not stable, is given in [14]. Many examples of stresses and strains in discretized bodies were given in [12, 15, 16, 18] .

11. Final Remarks

The equations of discretized bodies given in this report have been used as a starting point for further considerations, especially for detailed analysis of some special kinds of discretized bodies. The theory of discretized Cosserat bodies, in which each particle $d \in D$ can be interpreted as rigid body, has been studied in [12, 15] and applied to the problems of latticed structures in [3-11, 19]. Discretized multipolar and oriented bodies, in which each particle is either the set of free material points or one material point with a set of directors, have been discussed in [16]. More general approach to the mechanics of discretized bodies, in which we deal with non-elastic bodies, was given in [14]. In many special cases we are able to introduce the continuous model of the discretized body, governed by partial differential equations, [13]. The continuous model of discretized Cosserat media has been applied to some problems of lattice-type structures in [19].

The mechanics of discretized bodies is characterized by the simple and general form of the basic equations (2.4) and (3.3), by the resemblance of those equations to the known equations of micropolar elasticity and by the independence of the equations of the process of discretization. In the quasi-static problems the basic equations of mechanics of discretized bodies can be directly used as the equations of the known finite element method, [20], if we are to obtain the numerical solution of the problem.

Comparing the mechanics of discretized bodies to the continuum mechanics, we can conclude that all problems which can be solved using the equations of the continuum mechanics are not interesting as the problems of mechanics of discretized bodies. However, there are many problems of the theory of elasticity which cannot be solved; they are the problems in which we deal with various complicated boundary conditions, with a great number of concentrated forces and with non-continuous functions describing the properties of the body. In the theory of discretized bodies there are no boundary-value problems and all functions depend on the time coordinate only. Thus we can state that the equations of the mechanics of discretized bodies can be applied for rather complicated technical problems of non-smooth shells, shells and paltes made of many different materials and loaded by many concentrated forces, multiconnected bodies etc.

REFERENCES

[1] H. Frackiewicz: "Geometry of a discrete set of points" Arch. Mech. Stos., 3, 1966.

[2] H. Frackiewicz: "Mechanika osrodkow siatkowych (Mechanics of latticed media), in Polish, PWN, Warszawa 1971.

[3] M. Kleiber: "Lattice-type shells as the problem of elasticity of discretized bodies", Bull. Acad. Polon. Sci., sci. techn., 1972 (in press).

[4] M. Kleiber: "A consistent approximation in the linear theory of elastic lattice-type shells", Bull. Acad. Polon. Sci., sci. techn., 3, 1971.

[5] M. Kleiber, C. Wozniak: "On the equations of the linear elastic lattice shells", Bull. Acad. Polon. Sci., sci. techn., 3, 1971.

[6] M. Kleiber, Cz. Wozniak: "The equations of shallow lattice shells", Bull. Acad. Sci., sci. techn., 4, 1971.

[7] M. Kleiber, Cz. Wozniak: "The edge effect in the theory of lattice shells" Bull. Acad. Polon. Sci., sci. techn. 4, 1971.

[8] P. Klemm, S. Konieczny and Cz. Wozniak: "Dense elastic lattices of regular structure, Lattice shells" Bull. Ac. Pol. Sci., sci. techn., 11, 1970.

[9] P. Klemm and Cz. Wozniak: "Dense elastic lattices of regular structure", Bull. Acad. Polon. Sci., sci. techn., 10, 1970.

[10] P. Klemm and Cz. Wozniak: "Perforated circular plates under large deflections" Arch. Mech. Stos., 1, 1967.

[11] S. Konieczny and Cz. Wozniak: "Lattice-type structures as the problem of discrete elasticity", Bull. Acad. Polon. Sci., sci. techn., 12, 1971.

[12] Cz. Wozniak and P. Klemm: "Non-linear theory of the discrete Cosserat media", Bull. Acad. Polon. Sci., sci. techn., 1972 (in press).

[13] Cz. Wozniak: "Discrete Elasticity" Arch. Mech. Stos. 6, 1971.

[14] Cz. Wozniak: "Basic concepts of mechanics of discretized bodies", Arch. Mech. Stos., 3-4, 1972.

[15] Cz. Wozniak: "Discrete elastic Cosserat media" Arch. Mech. Stos., 1, 1973 (in press).

[16] Cz. Wozniak: "Equations of motion and laws of conservation in the discrete elasticity", 1, 1973 (in press).

[17] Cz. Wozniak: "Basic concepts of difference geometry" Ann. Polon. Math., 1972 (in press).

[18] Cz. Wozniak: "Theory of variated states in discrete elasticity" Bull. Ac. Pol. Sci., sci. techn., 1972.

[19] Cz. Wozniak: "Siatkowe dzwigary powierzchniowe (latticed shells and plates)" in Polish, PWN, Warszawa, 1970.

[20] O. C. Zienkiewicz: "The finite element method", Mc Graw-Hill, London, 1967.

Printed in the United States
By Bookmasters

Printed in the United States
By Bookmasters